FATIGUE CONSIDERATION in MACHINE DESIGN
ESSENTIALS of ENDURANCE ANALYSES

ROBERT E. LITTLE, Ph.D.
University of Michigan
Dearborn, Michigan

Fatigue Consideration in Machine Design

DEStech Publications, Inc.
439 North Duke Street
Lancaster, Pennsylvania 17602 U.S.A.

Copyright © 2020 by DEStech Publications, Inc.
All rights reserved

No part of this publication may be reproduced, stored in a
retrieval system, or transmitted, in any form or by any means,
electronic, mechanical, photocopying, recording, or otherwise,
without the prior written permission of the publisher.

Printed in the United States of America
10 9 8 7 6 5 4 3 2 1

Main entry under title:
 Fatigue Consideration in Machine Design: Essentials of Endurance Analyses

A DEStech Publications book
Bibliography: p.
Includes index p. 183

Library of Congress Control Number: 2019953999
ISBN No. 978-1-60595-605-3

HOW TO ORDER THIS BOOK

BY PHONE: 877-500-4337 or 717-290-1660, 9AM–5PM Eastern Time

BY FAX: 717-509-6100

BY MAIL: Order Department
**DEStech Publications, Inc.
439 North Duke Street
Lancaster, PA 17602, U.S.A.**

BY CREDIT CARD: American Express, VISA, MasterCard, Discover

BY WWW SITE: http://www.destechpub.com

To my grandchildren:
Victoria Ashley, Isabella Maria, Demetrios Adrian, Sophia Victoria, James Robert John, Lucas Roy, Alyssa Barbara, Jaron Richard Roger, Elianna Brenda, Nicholas Richard, and Romie Josephine Judy Louina.

Table of Contents

Prologue   ix

Acknowledgment   xi

Chapter 1. Wöhler's Tests 1

   1.1. Appendix WA   3
   1.2. Appendix WB/Engineering   7
   1.3. Appendix WC   14
   1.4. Appendix WD   15

Chapter 2. The Goodman Diagram, The Modified Goodman Diagram, and The Modified-Modified Goodman Diagram 23

   2.1. Appendix A   27

Chapter 3. Stress Redistribution Under Cyclic Loading ... 29

Chapter 4. R.R. Moore Four-Point Rotating-Bending Fatigue Test Machine 33

Chapter 5. Fatigue Crack Initiation and Opening Mode Propagation 39

Chapter 6. Scatter of Replicate s_a-N Curves and Endurance Limit Datum Values 43

Chapter 7. Service Fatigue Failure Exhibits 47

Chapter 8. Miscellaneous Failure Exhibits 57

Chapter 9. Miscellaneous SEM Micrographs 67

Chapter 10. Fatigue Effects 75

Chapter 11. Cumulative Damage 81

Chapter 12. The Ugly Truth About Laboratory Fatigue Tests. 85

Chapter 13. Ugly s_a-N curves 91

Chapter 14. The Actual Location of the Knee in Rotating Bending s_a-N Curves for Mild Steel Specimens 95

Chapter 15. Axial-load Strain-controlled (Low-cycle) Finite-life Fatigue Tests on Round Mild Steel Specimens 101

Chapter 16. Strain-controlled Bending, Torsion, and Combined Bending and Torsion Endurance Limits of Round Mild Steel Specimens 105

Chapter 17. Generalized Fatigue Models 109

Chapter 18. The Endurance Limit of Round Mild Steel Specimens under Rotating Four-point Bending with Limited Values of a Superimposed Steady Torsional Moment 113

Chapter 19. Endurance Limit Notch Sensitivity of Round Mild Steel Specimens Under Axial-loading or Bending 119
    19.1.   Appendix A   125

Chapter 20. The Effect of Mean Stress on Load-controlled Axial-load Endurance Limits of Round Mild Steel Specimens **127**
   20.1. Appendix A   134
   20.2. Appendix B   135

Chapter 21. Axial-load Endurance Limits for Mild Steel Sheet and Plate Specimens **137**

Chapter 22. Torque-controlled Torsional Endurance Limits of Round Mild Steel Specimens **141**

Chapter 23. Fatigue Factor of Safety **143**

Chapter 24. Fatigue Remedies and Fatigue Redesign **151**
   24.1. Alternative Ways to Reduce the Fatigue Stress   153
   24.2. Alternative Ways to Increase the Endurance Limit   159

Chapter 25. Bolted Joints **171**
   25.1. Appendix A   175
   25.2. Appendix B   176
   25.3. The Coupling Re-visited   181

Index   183

Author Biography   189

Prologue

I was motivated to write this book because machine design textbooks are typically written by authors who have an engineering mechanics perspective rather than a mechanical metallurgy perspective. The latter perspective includes the *common sense requirement that no plastic deformation is acceptable in mechanical design,* vis., *the maximum value of the stress imposed in service operation must not exceed the cyclic yield strength of the component material*, which for cyclic softening mild steels, is always less than the traditional static yield strength. This means that any fatigue analysis that explicitly includes a plastic strain component must never be used in machine design.

Machine components must maintain their original dimensions throughout their entire lifetimes. If the imposed service stress ever exceeds the cyclic yield strength of a mild steel component, then excessive deformation is the mode of failure and fatigue is not an issue.

The practical effect of the common sense constraint that the maximum *imposed service stress must never exceed the cyclic yield strength* of the material effectively dictates that machine design fatigue analyses pertain to very long fatigue lives, say about 10^6 to 10^8 alternating stress cycles. Given a mild steel, viz., a steel with an ultimate tensile strength less that about 180 to 200 ksi, this common sense constraint typically limits machine design fatigue analyses to pertain only to *endurance limits.* (See topic *Wöhler's Fatigue Tests*, in which the concept of an endurance limit for a mild steel is discussed.) Unfortunately, existing endurance limit data exhibit marked variability.

The fatigue literature does not provide sufficient data to establish credible machine design fatigue analyses for (i) shorter (so-called finite) fatigue lives or for (ii) high strength steels and other metals. Moreover, the stress-time history imposed in a laboratory fatigue test is overly simplistic compared to the actual (very complex) stress-time history imposed on a machine component in service operation. ***This problem has no solution.*** **Thus every fatigue design analysis must ultimately be justified or discredited by actual service performance.**

Perspective: Machine design is typically a process of first recognizing an analogous application that has exhibited satisfactory performance in service and then employing its estimated service performance loads and sizes to establish the sizes of the machine components of interest given their estimated loads. The metric for this extrapolation process is typically a naive factor of safety.

R. E. LITTLE

Acknowledgment

I would like to express my sincere appreciation to a former student, Waseem Youkhana, who took several of the photographs used in this book. Thank you, Waseem.

R. E. LITTLE

CHAPTER 1

Wöhler's Tests

AUGUST Wöhler was *the pioneer fatigue researcher* whose test results underlie much of the methodology presently employed in designing machine components to resist fatigue failure. As chief locomotive superintendent of the Royal Lower Silesian Railway (Prussia) he designed and built several different fatigue test machines starting about 1855 and subsequently conducted axial-load, rotating-bending, plane-bending and torsional fatigue tests. Perhaps his best known test results pertain to *comparative* rotating-bending tests on press-fit cantilever specimens machined from (i) wrought iron axles supplied in 1857 by the Phönix Company and (ii) cast steel axles supplied in 1862 by Krupp. As evidenced below, these tests clearly demonstrated that cast steel axles were superior to wrought iron axles.

Nevertheless, the fatigue data for wrought iron specimens are of much more interest in machine design. In particular, the cantilever rotating-bending fatigue test on the wrought iron specimen that endured 132,250,000 stress cycles without failing subsequently led to the conclusion that an *endurance limit* exists such that this specimen would not fail no matter how long the test were continued, viz., it would exhibit an "infinite" fatigue life. These test data also led to the concept of a distinct knee in the s_a-N curve that separates the finite life portion of the s_a-N curve from the "infinite life" portion pertaining to the *endurance limit*. Thus when it is evident during a given fatigue test that the number of imposed stress cycles exceeds the number of stress cycles associated with the knee in the s_a-N curve, the test can be terminated. Then the actual number of fatigue cycles imposed on the specimen without failure is plotted as a "datum value", but with an arrow pointing to the right which indicates the test specimen did not fail (DNF) and is subsequently referred to as a *run-out*.

TABLE 1.1. Wrought Iron Specimens

Value of the Maximum Bending Stress Imposed stated in centners per zoll^2	Number of Revolutions Made by the Test Specimen Before Fatigue Failure
320	56,430
300	99,000
280	183,145
260	479,490
240	909,810
220	3,632,588
200	4,917,992
180	19,186,791
160	132,250,000 Did Not Fail

A sketch of Wöhler's dual-cantilever rotating-bending fatigue test machine appears in **Appendix WA,** Figure WA1, along with axle specimen geometries C and D. In turn, Figure WA2 depicts the history of the normal stress due to bending as the specimen rotates. Figure WA3 presents alternative plots of the s_a-N curve for Wöhler's rotating bending fatigue data for geometry C specimens machined from wrought iron axles. Note that the number of alternating stress cycles pertaining to the respective knees in these two s_a-N curves *differ* because of the different s_a metrics employed in plotting.

Wöhler's display of wrought iron axle specimen fatigue failures at the 1867 Paris World's Fair was reported in *Engineering* (August 23, 1867) with sketches of the surface appearance of the fatigue failures with por-

TABLE 1.2. Cast Steel Specimens

Value of the Maximum Bending Stress Imposed stated in centners per zoll^2	Number of Revolutions Made by the Test Specimen Before Fatigue Failure
420	55,100
360	127,775
340	797,525
320	642,675
320	1,665,580
300	4,163,375
300	45,050,640

tions of the circumference of the enlarged tapered press-fit specimen end *deliberately* removed by machining prior to testing. See **Appendix WB**. These tests clearly illustrate the growth of fatigue cracks that (1) *initiate* at the fillet blending the cylindrical specimen diameter to its enlarged tapered press-fit specimen end and then (2) *propagate* inward across the axle cross section until (3) *abrupt fracture* occurs. Wöhler's display also included an axle test specimen that failed in *fretting fatigue* at its press-fit grip, viz., fatigue failure whose initiation is caused by fretting. In turn, **Appendix WC** presents a fatigue failure whose initiation was caused by corrosion, viz., *corrosion fatigue*. (It also provides an important lesson about learning.)

1.1. APPENDIX WA

Wöhler's plane-bending and axial-load fatigue test data established the fact that it is the *range of cyclic stress* that governs fatigue initiation rather than the maximum value of the cyclic stress. This revelation

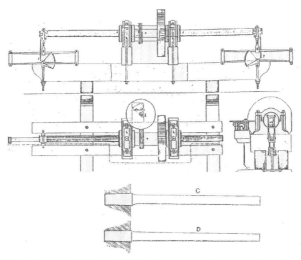

Figure WA1. *Top:* A sketch of Wöhler's dual-cantilever rotating bending machine. *Bottom:* Wöhler's press-fit wrought iron and cast steel axle specimens: Axle specimen geometry C with a fillet adjacent to its enlarged tapered press-fit end was used to demonstrate that the fatigue resistance of cast steel specimens is greater than that of wrought iron specimens. Wöhler also ran similar tests using test specimen geometry D with a sharp corner instead of a fillet and observed that both wrought iron and cast steel axle specimens exhibited markedly reduced fatigue lives. These tests established the most important fatigue design principle: *Always select the machine component geometry with the smallest practical stress concentration (factor).*

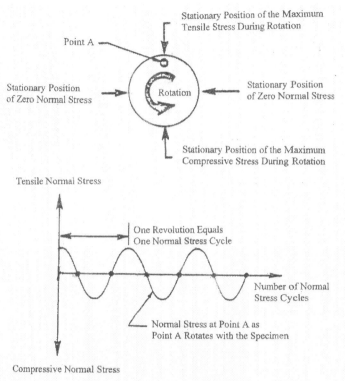

Figure WA2. *Top:* The rotation of Wöhler's axle specimen under the action of a constant bending moment generates a cyclic *alternating* (fully reversed) *normal stress* s_a at point A as point A rotates with the axle specimen. Each complete revolution generates one alternating normal stress cycle. *Bottom:* Plot of the sinusoidal alternating normal stress variation as the axle specimen rotates. Given a specific value for the *alternating* (fully reversed) *normal stress* s_a, the number of alternating normal stress cycles to failure is termed the *fatigue life of the failed axle specimen.*

led to the development of the Goodman diagram which attempts to describe endurance limits quantitatively. But I warn you in advance of my topic "The Goodman Diagram, The Modified Goodman Diagram, and The Modified Modified Goodman Diagram" that the Goodman Diagram itself is absolute nonsense, and its first two modifications only reduce the nonsense, but do not completely eliminate it.

Wöhler's tests on wrought iron axle geometry's C and D in Figure WA1 demonstrate clearly that the larger the fillet radius, (the smaller the stress concentration and) the greater the endurance limit. Avoiding square corners and employing fillets with large radii are perhaps the two most important concepts in designing against fatigue failure.

The s_a-N curve in Figure WA3 is plotted in the traditional manner, interchanging the independent and dependent variables. However, in my notation, s_a (lower case) connotes the numerical value of the amplitude of the imposed alternating stress, the independent variable; and N (capital) connotes the realization value of the *random* test response, the so-called fatigue life. The range of this fatigue life is so large that a logarithmic scale is always employed along the abscissa. Accordingly, I use the notation s_a-$\log_e[\mathit{fnc(pf)}]$ for the s_a-N curve, where *pf* is a specific probability of fatigue, *fnc* is the number of fatigue cycles, and s_a is the alternating stress amplitude.

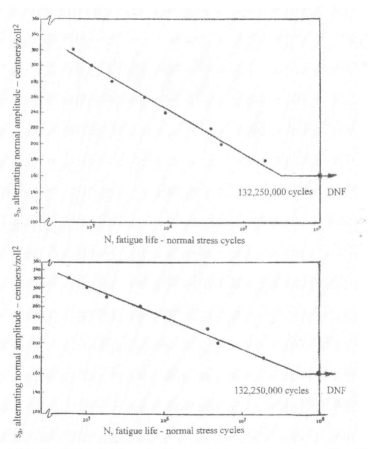

Figure WA3. *Top*: Wöhler's fatigue for his wrought iron axle specimens with a fillet plotted as a s_a-N curve with a *linear* alternating normal stress s_a metric-centners/zoll². *Bottom:* Wöhler's fatigue for his wrought iron axle specimens with a fillet plotted as a s_a-N curve with a *logarithmic* alternating normal stress s_a metric-centners/zoll².

My exemplar statistical analysis of Wöhler's rotating-bending s_a-N data for wrought iron specimens in **Appendix WD** employs a s_a-$\log_e[fnc(pf)]$ fatigue model with two straight-line segments. The horizontal segment's model is based on the premise that each individual test specimen has its own threshold value for fatigue crack initiation and that fatigue failure will occur for this specimen only if its threshold value is exceeded by the value for s_a imposed during the fatigue test. This threshold value concept leads to the concept of statistical distribution for the endurance limit that is analogous to the concept of a fatigue strength distribution pertaining to any specific number of fatigue cycles *fnc* of interest. The traces of the *medians* of these two statistical distributions are the two straight-line segments of the s_a-$\log_e[fnc(pf)]$ fatigue model: one fatigue strength segment for *finite life* and one endurance limit segment for *infinite life*. This s_a-$\log_e[fnc(pf)]$ fatigue model with two straight-line segments pertains to what I call mild steels, viz., steels with ultimate tensile strengths less than about 180 to 200 ksi.

The statistical literature does not provide an analysis for fatigue data with runouts. Rather it is always presumed in statistical analysis that a specimen will eventually fail if the test is run long enough. Accordingly, the statistical literature presumes that the test is merely suspended. When all test suspensions occur at the *same predetermined suspension fnc value* in a planned experiment test program, it is called *Type I* censoring. In turn, given bias-corrected maximum likelihood estimates of the actual parameter values for the presumed analytical model, the test program can be simulated again and again, say 30,000 times. Accordingly, there are two types of test outcomes:

(1) the respective test program imposed s_a values are so large that all specimens failed at fatigue lives smaller than the selected fatigue life suspension value, so that it is unlikely that any Type I suspensions would occur under continued testing with these s_a values: ***No Type I Censoring (Complete Data);***

(2) The respective test program imposed s_a values were such that one or more *Type I* suspensions occurred during the test program, so that it is likely that additional *Type I* suspensions would occur under continued testing with these s_a values:
Actual Type I Censoring.

However, I categorically assert that it is impossible to design for finite fatigue life (except in a very few special cases). Accordingly, our only option is to design using an estimated (median) *endurance limit* and a *naive factor of safety* (forthcoming).

1.2. APPENDIX WB/ENGINEERING

1.2.1. "Wöhler's Experiments on the Strength Of Metals" Engineering (August 23, 1867)

At the railway workshops belonging to the Niederschlesisch-Märkische Eisenbalm, in Prussia, M. Wöhler, the chief engineer, has for some years past been occupied in making a series of most interesting experiments on the strength of iron and steel when experiments on the strength of iron and steel when exposed to repeated strains within fixed limits and for a length of time. It is known, through the experiments of M. W. Fairbairn and of others, that the repeated application and removal of a load which is considerably below the breaking weight of any metallic bar will, after a number of such repeated applications, cause the fracture of the bar, and this apparent anomaly has been called the fatigue of metals. Exact laws for this so-called fatigue have, however, as yet never been obtained, and the few experiments made hitherto on this important subject have been utterly insufficient to derive any theoretical conclusions from them. Mr. Wöhler, having been charged with some comparative trials for ascertaining the relative value of steel and iron for railway axles, which are more than any other similar article exposed to repeated strains and to the effects of fatigue, decided upon commencing an independent series of experiments on a sufficiently large scale in order to arrive at some clear and instructive results. He constructed for this purpose a set of very ingenious machines which allowed him to expose his bars to vibrating actions and repeated strains within adjustable limits, and to observe carefully all the facts connected with each experiment. The results of M. Wöhler's numerous experiments and sketches of the special machines used by him are now exhibited at Paris by the directors of the Prussian State Railway, and we consider this exhibit one of the most interesting and important in the whole group of engineering objects collected there. The specimens, which are exhibited in a very small case in the Prussian machinery gallery, and are to be found only upon a careful search or inquiry, consist of a set of ten fractured bars of steel and of iron, and a small pamphlet affording the necessary explanations. From this pamphlet we have copied the sketches of the experimental machines used by M. Wöhler and the data of the strains borne by the different specimens, M. Wöhler, commenced his experiments with a machine (see diagram, Fig. 1) constructed for straining a cylindrical bar in a manner similar to that in which a railway axle is acted upon by the load it carries. The bar to be tested is fixed to the end of a rotating steel spindle *(a)*, supported in a pair of bearings *(b,b)*, and provided with a pulley *(c)* for obtaining rotation by means of a strap. The projecting end of the spindle *(d)* has a central conical recess into which one end of the bar is pressed, the other end being turned down into a small trunnion which runs in the bearing *(e)*, and upon which the adjustable action of the spring *(g)* exerts a constant downward pull. The bar is by this action bent down at the end to a fixed distance, and in its rotation the action of the spring makes this end take place in all directions successively all around the circumference of the bar. Taking the section through any one diameter, we find that in each rotation the bar is bent and unbent in two opposite directions, so for each revolution there is a strain exerted upon it in tension and in compression alternately. The amount of this strain is adjustable by the tension of the spring *(g)*, and measurable from the latter. The strain shown by the dynameter being effected in tension and in compression, it follows that the effect upon each fibre of the bar is equal to the sum of the two actions, or double the effect of the simple strain indicated by the spring. The breaking strain for bars of best fibrous iron under these conditions has been found to be no higher than 8 to 9 tons on the square inch, and the best soft cast steel had no more than 12 to 15 tons of tensile strength when strained in this manner, that is, strained between the limits of, say, 15 tons in tension and 15 tons compression alternately. There is, as an example of this kind, a broken bar of fibrous iron exhibited which had sustained nineteen millions of such alternate tensions and compressions with a strain of 9 tons on the square inch of section before fracture ensued. Now, if we take the total difference of change from a compression of 9 tons to a tension of the same amount, we find this to be 18 tons, or very nearly the tenacity of this material when strained in one direction only. In order to compare the effect of these strains with repeated strains which are of the same character, another machine represented by diagram Fig. 2 was designed by M. Wöhler. The bar under test (A) is in this case held upon two supports or linear bearings formed as knife-edges *(a,b)*, and the strain is transmitted to it at the centre by the bar *(m)* moved by a crank *(k)*. The spring *(f)*, adjustable in the amount of its tension, is set so as to balance the bending power effected by the crank through the leverage of the arms *(c,e)* and *(e,d)*. The bar is in this apparatus bent once for each turn of the crank, and the fibres are strained in each section in the same direction each time. The tension upon each fibre is therefore varying between the limits of zero and the strain indicated by the dynamometer. Under such circumstances fibrous iron

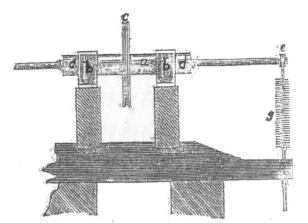

Figure 1.1.

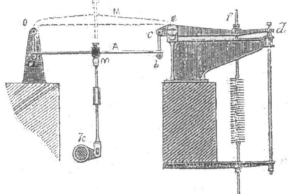

Figure 1.2.

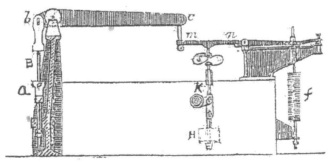

Figure 1.3.

Figure 1.4.

has been found to break with a tensile strain of 15 to 18 tons, and soft steel broke with strains of between 22½ to 25 tons on the square inch. This having been ascertained, it was desirable to find the amount of strain which such bars would bear if the strain applied in one direction was not taken off entirely, but only partly removed each time. The apparatus was for this purpose provided with the bar shown in dotted lines on Fig. 2 (M). On this bar an adjustable pin was set, so as to prevent the bar from unbending itself altogether when the pull of the crank was removed, and the latter obtained sufficient play in the bearing of the connecting rod to allow for this. The bar was then strained within given limits, both of which could be calculated from the indications of the tension spring at f. With this arrangement the remarkable fact was shown by experiment that a steel bar may be strained to 40 tons on the square inch, and released, say, to a tension of 20 tons with the same degree of safety as a tension varying between 12½ tons and 35 tons would give. A steel bar may therefore be loaded and unloaded between 20 and 40 tons strain with safety, while it will break down if the load of 40 tons is altogether removed each time, and it will equally break down with a load of only 15 tons when the latter is applied in opposite directions alternately. M. Wöhler has also employed an apparatus for testing the bars upon tension direct. This is represented in diagram, Fig. 3. The bar (B) is inserted between two gripping pieces (a,b), and the action of a crank (k) is transmitted to it by levers, the strain being counter-balanced by a spring, in the same manner as before. A second spring, of s form, is inserted for multiplying the length of movement between the test bar and the crank. The results obtained with this apparatus exactly coincide with those shown by the two other instruments, and the total result of their indications can be embodied in the following principle: viz., the fatigue of metals or their resistance to alternating and repeated strains does not correspond to the absolute amount of the strains repeatedly exerted upon them, but to the difference between the limits of variation of these strains, or to the dynamic effect of the changes which take place during the variation of the strain. The amount of total variation equal to the absolute breaking strength of fibrous iron has been found between 15 and 18 tons, and for the best soft cast steel it is between 22½ and 25 tons. Within the static limit of elasticity the same amount of variation is admissible with equal safety, no matter in what manner the total difference between the strains is arrived at. Alternating strains in opposite directions are acting in the same destructive sense, and must therefore be added to find the real difference or dynamic effect of the change. Besides these general laws, M. Wöhler has accidentally discovered several other facts of the greatest interest connected with these experiments. The bars occasionally inserted in his testing apparatus had a sudden change of section at one of their ends where they were inserted in the machine, a collar being turned on that spot, as shown by the sketch, Fig. 4. In all these cases the experiments proved that the strength of the bar at the spot where this sudden change of section takes place, is between one-fourth and one-third below the strength of the plain bar. The samples of fractured bars shown by M. Wöhler at the Paris Exhibition are some of the most characteristic specimens ever brought before the public. There is a set of such bars with their collars removed for a part of the circumference, and afterwards inserted into the testing-machine. The appearance of the fractures shows the effect of this in the most striking manner. We have in Figs. 5, 6, 7 and 8, reproduced the appearance of some of these fractures as nearly as this can be done, from a hasty sketch made on the spot, in the absence of any photographs or other good representations of these interesting specimens. Figs. 5, 6, and 7 show the fractures of bars with partly removed collars, and Fig. 8 shows a bar with a complete collar. The parts of the fracture shown dark have the appearance of fine soft steel of grey colour, while the rest shows a whitish very close-grained fracture and a wavy surface. It is perfectly obvious, from the appearance of these samples, that fracture took place gradually all around the collar at first, and the rest of the bar did not give way until the crevice or crack so formed had reached deep enough to the centre to reduce the section of the unbroken portion below the ultimate limit of strength.

Mr. Wöhler has also established the fact that a gradual change of section, such as produced by a round hollow of a sufficiently large diameter, does not cause such a weakening effect upon the steel bar as a right angle. The reason for this has been sought for in the mode of action of a cutting tool when working into a sharp corner, and it is plausible to think the cutting edge of the tool may have a similar cleaving action upon the solid steel as the corner of a diamond has upon a glass plate. According to this view, there would be a crevice or cleavage plane of infinite smallness produced in the corner between the collar and the rest of the bar, which weakens the section of the latter. It is, however, more probable that this effect is due to the break of regularity in the bending action exerted upon the fibres on the spot where two different sections meet abruptly. The appearance of the fractured surface next to the collar resembles very strikingly the fracture of those wedge-shaped pieces of steel which are forced out when a bar is broken crossways, and which represent the fractures through the compressed fibres. The darker portions in the centre of M.

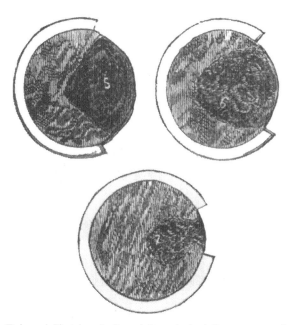

Figure WB1. Enlarged Sketches 5, 6, and 7, each depicting a removed portion of the so-called collar, viz., the increased diameter of the tapered press-fit end portion of Wöhler's axle test specimens. In each case the fatigue crack *initiated* at the fillet stress concentration between the long cylindrical portion of the specimen and its enlarged tapered press-fit end. Then each fatigue crack *propagated* inward until abrupt fracture occurred. The *abrupt fracture* portion of the failure surface is depicted by the dark portion of the sketch surrounding the numbers 5, 6, and 7. Note the relatively small area of the abrupt fracture for axle test specimen 7.

Wöhler's bars, on the contrary, resemble the fractures of bars ruptured by tension. It seems, therefore, that the sharp corner, by interfering with the flexibility of the bar, concentrates the action of the bending force upon a more limited space, which is therefore less capable to resist the effects of repeated strains or vibrations. This seems to us a very important fact, when considering the nature of the strains to which iron and steel often are exposed in practice. It is possible that further inquiry into this subject may at some not very distant time prove that the method of superposing cylindrical rings in the construction of ordnance, which cause a series of such sudden changes of section, is a source of weakness and danger, since it interferes with the freedom of vibration in the whole mass of the gun, and caused places of weakness at each break of section. There are

Figure WB2. Enlarged Sketches 8 and 9. Sketch 8 is a continuation of Sketches 5, 6, and 7, but with a so-called collar that is complete. It depicts the fatigue crack initiating at the fillet around the entire circumference of the so-called collar and then propagating inward until abrupt fracture occurred. I deleted the shading of the collar in the article so that it is clear that Sketch 8 actually depicts a typical fatigue failure of the axle specimen, the same as Sketch 10 which follows. On the other hand, Sketch 9 in turn depicts a fretting fatigue failure where two main fatigue cracks originated at opposite sides within the axle specimen press-fit region because of local fretting and then each of these main *fretting fatigue* cracks propagated inward from its respective side until abrupt fracture occurred along the axle specimen center. The shape of the abrupt fracture region indicates that this specimen still had a slight "wobble" (runout) during testing. (Wöhler pressed a bar against the bearing end of press-fit axle specimens to eliminate (reduce) the wobble (runout) caused by the press-fit axle specimen not being perfectly aligned (concentric) with the tapered hole in the end of the rotating center drive shaft.)

Figure WB3. Enlarged Sketches 10 and 11. Sketch 10 depicts a typical fatigue failure fracture appearance. (This sketch is what Sketch 8 in the article should have looked like.) I removed the shading of the so-called collar in Sketch 11 because it makes no sense. However, if the so-called collar were left blank, as in Sketch 10, then the resulting fatigue failure appearance indicates that a very large alternating stress was imposed on an axle specimen, resulting in several fatigue cracks initiating almost simultaneously at different *locations* and distinct *elevations* on the fillet around the circumference of the axle specimen. Then, as the respective fatigue cracks propagated, their respective fatigue crack front elevations diminished and converged to form a single crack front in subsequent propagation. The vertical junctions between these different elevations are short radial lines when looking directly at the fracture surface. However, depending on the light source, these vertical junctions cracks can cast shadows that highlight the different elevations.

some more facts, which seem to point in the same direction, to be gather from M. Wöhler's experiments. Fig. 9 shows a steel bar broken across its largest section in consequence of the weakening influence produced at that place by the pressure of the end of the spindle in the apparatus to which it was fastened. It has been found, particularly with regard to steel, that the strength of the bar is reduced by about one-third through the influence of a very moderate pressure exerted upon that bar by the conical hole in the centre of the mandril of the machine (Fig. 1), into which the end of the bar is jammed. This is an analogous case to the shrinking of rings upon each other by building up large pieces of ordnance; and considering the importance of all questions connected with this branch of engineering, we can only recommend the repetition of experiments similar to those of M. Wöhler, and the most earnest investigation of this as yet almost unknown subject, viz., the laws of the fatigue of metals. We have in Figs. 10 and 11 shown two sections of steel fractured by the apparatus (Fig. 1) of M. Wöhler under normal conditions. The radial shades in Fig. 11 are due to an unevenness of fracture at those spots, but they contain no cracks nor fissures, and are simply the result of the higher portions throwing shadows upon those which lie a short distance below. The fractures of fibrous iron which M. Wöhler has also exhibited at Paris show that this process of alternate bending has no tendency to change the nature of the metal and to induce crystallization. There is one specimen which was bent more than nineteen millions of times before it broke, and yet the fracture is as fibrous as the original bar from which the specimen was made. M. Wöhler's modest exhibition may have been overlooked by ninety-nine out of every hundred professional visitors to the Exhibition, yet we believe ourselves justified in saying that his scientific and patient experiments will be referred to long after the majority of those things which have drawn a shower of medals and ribbons upon themselves at present will be dismissed and forgotten.

1.3. APPENDIX WC

In case you may think that Wöhler solved the problem of failures of axles in the railroad industry, I quote the following from an article on fatigue by B. P. Haigh with unfortunate title "Brittle Fracture of Metals" published in *Engineering* (December 12, 1930).

"Fatigue failure might, it was found, be greatly hastened by chemical action. In 1916, the railways were troubled with axle failure on some new coaches. The load and the steel were the same as had proved satisfactory in previous years, but whilst the axles had a life of many years in the old stock, failures were then occurring in less than six months. The cause was ultimately discovered to lie in the fact that, in the new coaches, the discharge from the lavatories fell on the axles in question. When this was corrected there was no longer trouble with broken axles."

Remark One: Haigh's article also mentions his grandfather had recorded and analyzed failures of locomotive axles and by 1842 he had recognized that engine mileage was the governing factor in these axle failures.

Remark Two: Haigh evidently had tested only mild steels during his career and was unaware that not all metals have an endurance limit. I quote the following from his article.

> "Wöhler, also a locomotive engineer (as was Haigh's grandfather), commenced his classical experiments on repeated and alternating stresses, and arrived definitely at the idea that in all cases there was a safe limit to the range of stress within which fracture would not occur, no matter how many the repetitions. This view since became general, but it is now being challenged in America, on the basis of experiments on some aluminum alloys in which no such limit has yet been found. It would be very inconvenient should this view prove true, but, the speaker (Haigh) believed that the evidence, taken as a whole favored the older idea."

Haigh's problem is as old as mankind. No matter how much you think you know and no matter how much experience you may have, you do not learn by repeating what you have already done. Rather you learn when you do something different.

1.4. APPENDIX WD

A-Basis and B-Basis Statistical Tolerance Limits Based on a Conceptual s_a-$\log_e[fnc(pf)]$ Model with Two Straight-Line Segments for a Material With an Endurance Limit, Given Load-Controlled s_a-fnc Data with One or More Run-Outs and Presuming Identical Conceptual (Two-Parameter) Normal Fatigue Strength and Endurance Limit Distributions.

The primary objective of conducting a s_a-*fnc* experiment test program for a material with an endurance limit almost always is (or should be) to estimate $s_{endurance\ limit}$ (50). Moreover, since the small-sample up-and-down test method provides a statistically efficient and reasonably precise estimate of $s_{endurance\ limit}$(50), it is seldom statistically rational to run a conventional s_a-*fnc* test program. Nevertheless, s_a-*fnc* data still dominate the fatigue literature and thus warrant analysis.

The downward sloping straight-line finite-life segment of the two segment s_a-$\log_e[fnc(pf)]$ model for material with an endurance limit is estimated based on the presumption of a homoscedastic normal *strength* distribution, viz., given fatigue life datum values generated in an ordinary s_a-*fnc* test program the respective deviations of the plotted fatigue life datum values from the estimated median downward-sloping, straight-line finite-life segment are measured in the vertical direction (strength) rather than in the horizontal direction (life). This *conditional* ML analysis is consistent with the associated ML analysis for the horizontal "infinite-life" segment. Thus the presumptions (assertions) underlying the two-segment s_a-$\log_e[fnc(pf)]$ model are (i) its downward-sloping, straight-line, finite-life segment is actually the trace of the median of the alleged homoscedastic conceptual (two-parameter) normal *fatigue strength* distribution and (ii) its horizontal straight-line segment is actually the trace of the median homoscedastic conceptual (two-parameter) normal *endurance limit* distribution, and further (iii) these two conceptual normal distributions are *identical* at the knee. The actual value for the standard deviation of the homoscedastic conceptual (two-parameter) normal strength distribution is thereby equal to the actual value for the standard deviation of the conceptual (two-parameter) normal endurance limit distribution. Accordingly, when the actual value for the standard deviation of the presumed normal fatigue strength distribution along the downward-sloping, straight-line, finite-life segment of the two-segment s_a-$\log_e[fnc(pf)]$model has been estimated, this bias corrected estimate can subsequently be employed in the accompanying maximum likelihood analysis to estimate the actual value for the median of the presumed conceptual (two-parameter) normal endurance limit distribution. This procedure mitigates the problem that a reasonably precise estimate of the standard deviation of the presumed conceptual (two-parameter) normal endurance limit distribution is seldom available.

Microcomputer programs *SNM(J)ELTL*, $J = 1,2$, employ conditional maximum likelihood analyses for (only) the downward-sloping straight-line finite-life segment of the two-segment s_a-$\log_e[fnc(pf)]$ model. Then the respective *median-bias-corrected* estimates of the actual values for the standard deviations of the presumed normally distributed fatigue strength distributions are subsequently employed in accompanying maximum likelihood analysis to estimate the actual values for the respective median endurance limits. In turn, the intersection of the median downward-sloping, straight-line, finite-life segment of the two-segment s_a-$\log_e[fnc(pf)]$ model with its median horizontal straight-line endurance-limit segment establishes the value for ML est*(fnc$_{knee}$)*. Then, given this ML est*(fnc$_{knee}$)*value, the original finite-life conditional maximum likelihood analysis is extended by using Student's non-central t to compute *A*-basis and *B*-basis statistical tolerance limits that allegedly bound the actual values for the first and tenth percentiles of the presumed identical fatigue strength and endurance limit distributions.

The Box-Cox test statistic is employed to decide statistically when it is appropriate to switch from a linear s_a metric to a logarithmic s_a metric. Given Wöhler's exemplar data for wrought iron, this test statistic realization value is equal to $21.9066 - 20.1799 = 1.7267$. Accordingly, although this value clearly favors employing a logarithmic alternating stress amplitude metric, it is not large enough to cause rejection of the null hypothesis the linear s_a metric is correct (given an acceptable probability of incorrectly rejecting this null hypothesis is equal to 0.05). Accordingly, we opt to employ a linear s_a metric. Note that the use of a linear s_a metric also provides a more conservative (safer) value.

Discussion: Unfortunately the common practice in s_a-*fnc* testing is to let a combination of the location of (only) the finite-life straight-line segment of the two-segment s_a-$\log_e[fnc(pf)]$model and some alleged knowledge of the location of its knee establish the "estimated" value for the median endurance limit. This practice is particularly inept because the actual location of the knee is not fixed. It depends on a number of factors, e.g., the value for the tensile mean stress, the specimen size and shape (especially if hollow), the surface finish, the mode of loading, etc. It can vary from about 10^4 to about 5 times 10^7 cycles for plain-carbon and alloy steels with ultimate tensile strengths up to about 180 to 200 ksi.

TABLE 1.3.

C> Type Eltldata		
8		Number of Different Stress Amplitudes with Failures
1		Number of Replicate Tests at First Stress Amplitude
320	56430	First Stress Amplitude and Fatigue Life in Cycles
1		Number of Replicate Tests at Second Stress Amplitude, et cetera
300	99000	
1		
280	183140	
1		
260	479490	
1		
240	909810	
1		
220	3632590	
1		
200	4917990	
1		
180	19186790	
1		Number of Stress Amplitudes with Run-Outs
1		Number of Replicate Tests at First Stress Amplitude
160	132250000	First Stress Amplitude and Fixed Number of Run-Out Cycles, et cetera
95		scp = 95 (Integer Value) for A-basis and B-basis Statistical Tolerance Limits
10		pf in Per Cent (Integer Value): A-basis= 01, B-basis = 10
30000		Number of "Replicate" Pseudorandom Data Sets Used in Simulation (30000 Maximum)
185 381 355		A New Set of Three Three-Digit Odd Seed Numbers

TABLE 1.4.

C> Copy Eltldata Data
1 files(s) copied
C> SNM1ELTL
Given s_a-*fnc* datum values with one or more run-outs and presuming a two-segment s_a-$\log_e[fnc(pf)]$ model in a conditional ML analysis with the homoscedastic conceptual (two-parameter) normal fatigue strength distribution augmented parameterization

$$y = \{s_a - clp0 - clp1 \cdot [\log_e(fnc)]\}/csp$$

for its downward-sloping straight-line finite-life segment and with Version LS statistical bias-correction factors and the exact multiplicative median

(continued)

TABLE 1.4. continued

statistical bias-correction factor for generic est(csp)		
fnc	s_a	est[s_{fs}(50)]
56430	320	312.5
99000	300	299.2
183140	280	284.6
479490	260	261.8
909810	240	246.7
3632590	220	213.9
4917990	200	206.7
19186790	180	174.5

Estimate of the Actual Value for the Median Endurance Limit
170.0

Number of Fatigue Cycles at Which est[s_{fs}(50)] is Equal to the Estimated Median Endurance Limit
23186401

Lower 95% (one-Sided) Statistical Tolerance Limit that Allegedly Bounds the Actual Value for s_{fs}(10) at 23186401 Cycles Computed Using Students Non-Central t
149.3

Lower 95% (one-Sided) Statistical Tolerance Limit that Allegedly Bounds the Actual Value for s_{fs}(10) at 23186401 Cycles
149.4

Based on the 95th Percentile of the Pragmatic Sampling Distribution Comprised of 30000 "Replicate" Realizations Values for the Lower 95% (one-Sided) Statistical Tolerance Limit that Allegedly Bounds the Actual Value for s_{fs}(10) at 23186401 Cycles Computed Using Students Non-Central t The central 100(p)% of this translated pragmatic sampling distribution is bounded by the following values:

25%	75%
144.0	154.5
12.5%	87.5%
139.7	158.1
5%	95%
135.5	161.4
2.5%	97.5%
132.6	163.5
0.5%	99.5%
127.2	168.1

Box-Cox T(1) test statistic realization value= .219066D+02

Number of *Type I* Censored Datum Values	Number of "Replicate" Pseudorandom Data Sets
0	1859
1	26319
2	1822

TABLE 1.5.

C> Copy Eltldata Data
1 files(s) copied
C> SNM2ELTL
Given s_a-*fnc* datum values with one or more run-outs and presuming a two-segment s_a-$\log_e[fnc(pf)]$ model in a conditional ML analysis with the homoscedastic conceptual (two-parameter) nonnal fatigue strength distribution augmented parameterization
$$y = \{\log_e (s_a) - clp0 - clp1 \cdot [\log_e(fnc)]\}/csp$$
for its downward sloping straight line finite-life segment and with Version LS statistical bias-correction factors and the exact multiplicative median statistical bias-correction factor for generic est(csp)

fnc	s_a	est[$s_{fs}(50)$]
56430	320	317.6
99000	300	300.7
183140	280	283.3
479490	260	258.0
909810	240	242.4
3632590	220	211.8
4917990	200	205.7
19186790	180	180.2

Estimate of the Actual Value for the Median Endurance Limit
169.7

Number of Fatigue Cycles at Which est[$s_{fs}(50)$] is Equal to the Estimated Median Endurance Limit
35539153

Lower 95% (one-Sided) Statistical Tolerance Limit that Allegedly Bounds the Actual Value for $s_{fs}(10)$ at 35539153 Cycles Computed Using Students Non-Central t
158.2

Lower 95% (one Sided) Statistical Tolerance Limit that Allegedly Bounds the Actual Value for $s_{fs}(10)$ at 35539153 Cycles
158.2

Based on the 95th Percentile of the Pragmatic Sampling Distribution Comprised of 30000 "Replicate" Realizations Values for the Lower 95% (one-Sided) Statistical Tolerance Limit that Allegedly Bounds the Actual Value for $s_{fs}(l0)$ at 35539153 Cycles Computed Using Students Non Central t

The central 100(p)% of this translated pragmatic sampling distribution is bounded by the following values:

25%	75%
155.3	161.0
12.5%	87.5%
153.1	163.1

(continued)

TABLE 1.5. continued

5%	95%
150.9	165.0
2.5%	97.5%
149.5	166.3
0.5%	99.5%
146.9	168.9
Box-Cox T(0) test statistic realization value = .201799D+02	
Number of *Type I* Censored Datum Values	Number of "Replicate" Pseudorandom Data Sets
0	107
1	29789
2	104

Postscript to my use of alternative metrics in my endurance limit analyses:

When I am occasionally asked to audit an engineering analysis (calculation), whatever the presumed mode of failure, I am almost always disappointed to find a single set of presumptions employed (adopted) in analysis. Typically there are a few alternative presumptions that are either equally credible or at least, reasonably credible. In fact, the only engineering analyses that have no alternative solutions are found in textbooks or in professional codes.

CHAPTER 2

The Goodman Diagram, The Modified Goodman Diagram, and The Modified-Modified Goodman Diagram

WÖHLER concluded, based on his pioneer fatigue tests, that it is the range of cyclic normal stress rather than the maximum value of the cyclic normal stress that establishes the value for the endurance limit. Goodman, a civil engineer, attempted to explain this endurance limit behavior using a model based on the assertions that (i) a cyclic normal stress is actually a *dynamic* normal stress; (ii) thus the value for Wöhler's cyclic normal stress must be increased by a factor of two; and (iii) the endurance limit is reached when the maximum value of this dynamic normal cyclic normal stress is equal to the ultimate tensile strength. Accordingly, Goodman determined that the endurance limit is equal to one-third of the ultimate tensile strength (which remarkably was approximately equal to Wöhler's ratio). In turn, Goodman extended his model to include the effect of superimposed mean normal stress.

By about 1900 it was common knowledge among civil engineers that a zero velocity impact is actually a dynamic load and its value is equal to two g's, one g to stop the acceleration of the dropped weight (mass) due to gravity and one g merely to support the (now static) dropped weight (mass). Accordingly, Goodman asserted that Wöhler's cyclic normal stress with a range equal to two thirds of the ultimate tensile strength is actually a dynamic cyclic normal stress with a range of four-thirds of the ultimate tensile strength. Thus, given Wöhler's cyclic normal stress with a mean value equal to zero, the minimum value for Wöhler's cyclic normal stress must be equal to a minus one-third of the ultimate tensile strength. See Figure G1, where Goodman's maximum value for Wöhler's cyclic normal stress is equal to the ultimate tensile strength regardless of the value for the mean stress. Note that, for example, given a nominal mean stress equal to one-half of the ultimate tensile strength, the range of Wöhler's normal stress associated with the endurance limit is reduced by one-half.

Figure G2 presents the first modification the Goodman's diagram. Figure G3 in turn presents the second modification Goodman's diagram.

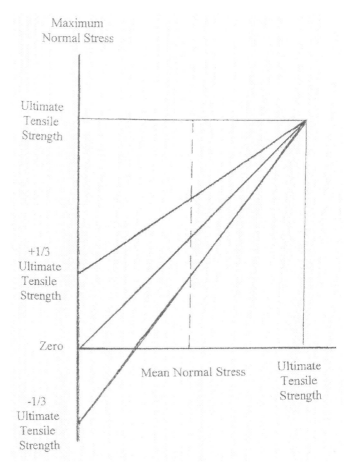

Figure G1. The original Goodman diagram. Suppose the imposed mean normal stress (plotted along the abscissa) is one-half the ultimate tensile strength. Then this Goodman diagram asserts that the endurance limit is one-sixth of the ultimate tensile strength and that the corresponding minimum and maximum cyclic normal stresses are respectively: one-third and two-thirds of the ultimate tensile strength. Specifically, the *amplitude of the alternating cyclic normal stress* is one-sixth of the ultimate tensile strength and, *by definition,* is equal to the endurance limit.

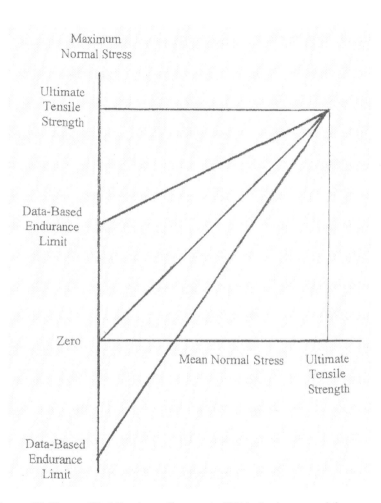

Figure G2. The modified Goodman diagram. As Wöhler's pioneer work became more widely known, several fatigue investigators built laboratory fatigue test machines and conducted tests on mild steel specimens. Their results generated a consensus that the endurance limit of mild steels was markedly greater than one-third of the ultimate tensile strength. According the Goodman diagram was modified by replacing one-third of the ultimate tensile strength by a *data-based* value for the endurance limit, typically about one-half of the ultimate tensile strength.

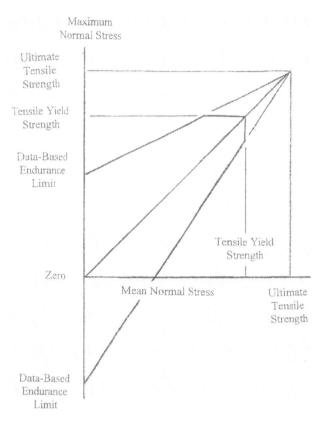

Figure G3. The modified-modified Goodman diagram. A German industrial standard was published in 1933 that limited the maximum value of the normal cyclic stress to the (static) tensile yield strength. This *common sense* limit was long overdue at that time. Nevertheless, starting in the 1960's, most fatigue investigators ran strain-controlled fatigue tests at amplitudes so large that the resulting data has no machine design application.

Remark: You may think that, after two common sense practical modifications, the resulting modified-modified Goodman diagram is suitable for use in designing mild steel machine components. If so, you are wrong, but so are most, if not almost all authors of machine design textbooks. See topic "The Effect of Mean Stress on the Endurance Limit of Mild Steel Specimens."

2.1. APPENDIX A

The use of Haigh-Soderberg coordinates is preferable in machine design applications because the load lines associated with the definition of a naive factor of safety are more intuitive.

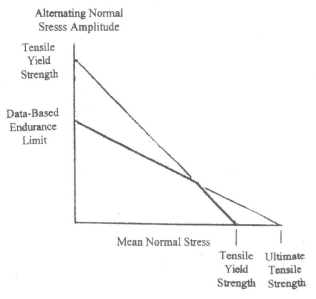

Figure AG1. The modified-modified Goodman diagram plotted on traditional Haigh-Soderberg coordinates.

Remark: I do not believe that plotting static strengths on an endurance limit diagram makes any sense, except when these strengths serve only as reference values being used in the numerical construction of the endurance limit mean stress diagram of interest. Thus, I always replace the static tensile yield strength by my estimated value for the cyclic yield strength at the same number of alternating normal stress cycles as pertains to the endurance limit. Then I can rationally assert that this cyclic yield strength locus is just as physically meaningful as the data-based endurance limit locus (even though I cannot estimate its value as precisely).

CHAPTER 3

Stress Redistribution Under Cyclic Loading

IN the late 1930's and the early 1940's numerous X-ray studies of the redistribution of stress under cyclic loading were published in Germany. (This was long before dislocation theory and acoustic emissions technology became established.) I first briefly present a little of that X-ray data just to document stress redistribution occurs during cyclic loading. Then I present some of my strain-gage-based data to show that stress redistribution can also be detected in less sophisticated ways.

Remark: I was first intrigued by X-ray stress analysis when I read about a test in which Glocker annealed a low-carbon steel tension specimen and then used an X-ray analysis to confirm that the tension specimen surface was actually stress-free. Then he loaded the tension specimen to about two-thirds to three-quarters of the tensile yield strength before unloading it. Upon subsequent X-ray analysis he found that the tension test specimen exhibited large compressive residual surface stresses. *Obviously the **surface** metal grains had yielded at a stress well below the static yield strength.* No mechanics or metallurgy book that I knew of had ever mentioned that *behavior!* I was left to rationalize the cause of this behavior myself. I decided the surface grains have fewer deformational constraints than interior grains and thus surface grains are less resistant to deformation (by slip) than are interior grains.

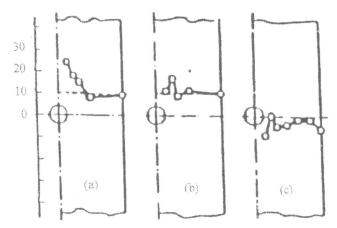

Figure R1. This figure displays X-ray stress data pertaining to axial-loading of a mild steel sheet specimen with a central hole. All cyclic stresses are below the cyclic yield strength. (a) X-ray stress distribution under a static nominal stress equal to 10 kg/mm². (b) After 16,200,000 cycles with a mean stress equal to 10 kg/mm² and an alternating stress equal to 5 kg/mm². (c) Completely unloaded.

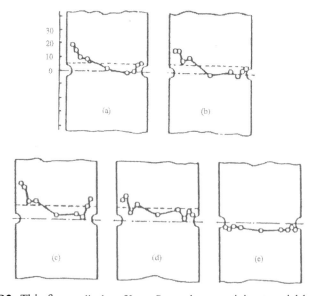

Figure R2. This figure displays X-ray Stress data pertaining to axial loading of a mild steel sheet specimen with double edge notches. All cyclic stresses are below the cyclic yield strength. (a) X-ray stress distribution under a static nominal stress equal to 6 kg/mm². (b) After 7,000,000 cycles with a mean stress equal to 6 kg/mm² and an alternating stress equal to 5 kg/mm². (c) Under a static nominal stress of 10 kg/mm². (d) After 7,700,000 cycles with a mean stress equal to 10 kg/mm² and an alternating stress equal to 5 kg/mm². (e) After unloading.

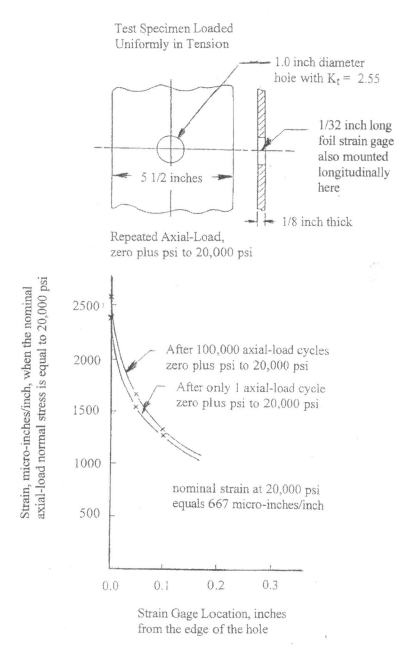

Figure R3. My strain-gage-based stress redistribution data. Massive yielding occurred upon initial axial-loading. The nominal yield strain was 1350 micro-inches/inch and the constrained yield point strain was 1700 micro-inches/inch. Then massive redistribution occurred during the first axial-load cycle. In fact, most of the redistribution occurred during the first 10 cycles.

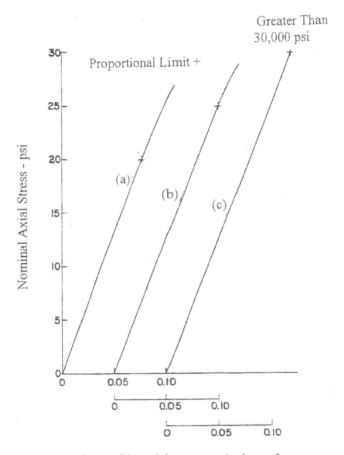

Figure R4. Load-Elongation plots for: (a), a statically-loaded previously untested fatigue specimen; (b), a fatigue runout at 10^7 cycles, previously tested with a nominal stress fluctuating from zero plus psi to 20,000 psi, and then statically loaded; (c), a fatigue runout at 10^7 cycles, tested with a nominal stress fluctuating from 10,000 psi to 30,000 psi, and then statically loaded. The axial load is expressed in terms of corresponding nominal normal tensile stress. The elongation ostensibly pertains to the entire specimen but actually is the load-induced axial movement of the lower grip relative to the upper grip. The fundamental issue is that cyclic-dependent behavior is (or can be) evident both "microscopically" and "macroscopically" in load-controlled fatigue tests.

CHAPTER 4

R.R. Moore Four-Point Rotating-Bending Fatigue Test Machine

MORE fatigue data has been generated using a R. R. Moore four-point rotating-bending fatigue machine than any other fatigue test machine. The reason is simple. It typically runs at about 10,000 RPM, viz., about five times faster than axial-load and crank-driven fatigue test machines. See Figure RRM1.

Figure RRM2 presents the s_a-N curve often found in a machine design textbook. It is allegedly based on a R. R. Moore test data for mild steel specimens. Note that the values pertaining to 10^3 cycles make no sense, except as a fictitious values employed to establish the slope of the finite-life portion of the s_a-N curve. Nevertheless, the finite life straight-line portion of the s_a-N curve can extend somewhat above the static yield strength without a noticeable change in slope because the test specimen is rotating too fast to observe any deformation. Accordingly, the obvious increase in the specimen temperature is the primary indicator that massive micro-plastic deformation is occurring during testing. It was demonstrated in the early 1920's that hysteresis loops can form during R. R. Moore rotating bending tests and cause stress and strain to be out of phase.

It is also fictitious that the knee in a s_a-N curve occurs at 10^6 stress cycles. See my fatigue topic "The Actual Location of the Knee in Rotating Bending s_a-N Curves for Mild Steel Specimens".

Micro-plastic behavior also occurs at stresses well below the yield strength and cause both the original linear specimen stress and strain distributions in a R. R. Moore test specimen to redistribute as illustrated in Figure RRM3 *(top)*. In contrast, the specimen strain distribution never changes during a crank-driven deflection-controlled plane-bending fatigue test, Figure RRM3 *(bottom)*. This difference in fatigue test control, constant-moment versus crank-driven constant-deflection,

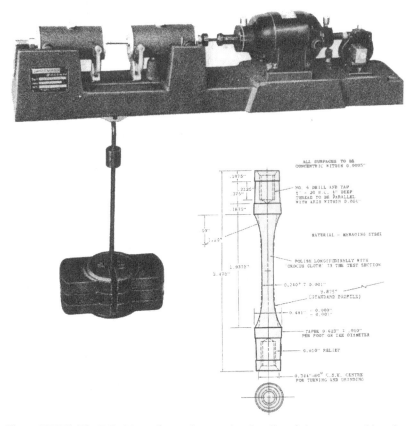

Figure RRM1. The R.R. Moore four-point rotating bending fatigue test machine features a uniform bending moment along the specimen. The magnitude of this uniform moment is established by the (fixed) value of the hanging weight. Its standard test specimen (light dashed lines) and one of its many modifications (dark solid lines). The diameter of the specimen at its throat is 0.3 inches for most mild steel specimens, but is smaller for higher ultimate tensile strength steels. The radius of the specimen throat is always large enough to presume a stress concentration factor is approximately equal to one.

is the reason why R.R. Moore rotating bending endurance limit data are somewhat lower than constant deflection plane-bending endurance limit data. See Figure RRM4.

Remark: Figure RRM5 presents a very extreme demonstration that the massive accumulation of micro-plastic behavior generates hysteresis loops. The imposed range of nominal normal strain is controlled in this (slow) axial-load test while the corresponding specimen nominal normal stress is also continually recorded. Although the first strain

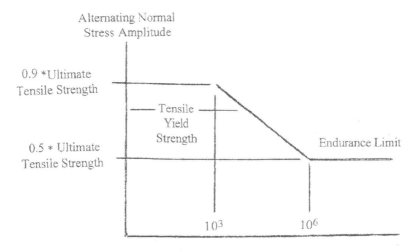

Figure RRM2. A typical machine design textbook plot of a s_a-N curve for mild steels. This representation is actually based on R. R. Moore rotating bending fatigue data, but is often *incorrectly* alleged to be a generalized s_a-N curve for mild steels. It has several serious problems. The actual location of the knee in a s_a-N curve with an endurance limit depends on several factors that are discussed elsewhere, depending on which the knee can occur between about 10^4 to about 5×10^7 alternating normal stress cycles. The omnibus times the ultimate tensile strength estimate of the endurance limit is not credible. It does not account for the well-established fact that the endurance limits of mild steels are also strongly influenced by the static tensile yield strength, viz., given two mild steels with the same ultimate strength, the one with the higher static tensile yield strength will (almost always) have the higher endurance limit.

cycle appears to be elastic, it is not. The micro-plastic behavior that occurs during this first cycle is merely too small to be detected by eye. However, it accumulates during each strain cycle and, in this very extreme example, a hysteresis loop is clearly evident at 10 cycles. Note that this hysteresis loop broadened markedly as the strain cycling continued. Note also that the corresponding range of stress decreased markedly. This behavior is called *cyclic softening*. The stress redistribution described in Figure RRM3 *(bottom)* is essentially a mini version of this extreme example.

Remark: One of the most interesting phenomena in fatigue pertains to Armco iron, which might be regarded as a mild steel. It is the only metal that I know of that has its rotating bending endurance limit higher than its *static* tensile yield strength.

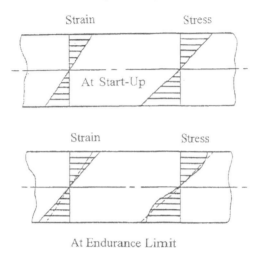

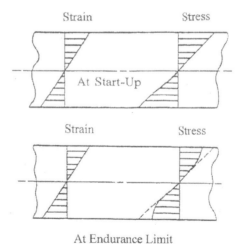

Figure RRM3. Stress redistribution during a constant-moment R. R. Moore rotating-bending fatigue test and during a crank-driven deflection-controlled plane-bending fatigue test, each test employing the same *cyclic-softening* mild steel. See Figure RRM5.

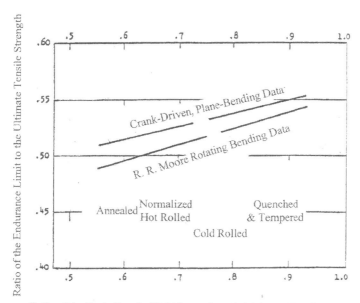

Ratio of the Static Tensile Yield Strength to the Ultimate Tensile Strength

Figure RRM4. A plot of what I call *generic design curves* for estimating the endurance limits of round mild steel fatigue specimens under rotating-bending and under crank-driven deflection-controlled plane-bending in the preliminary stage of *sizing* design stress analysis. This plot is based on a large compilation of published endurance limit data with each line lying below the majority of the respective plotted datum values. Once the preliminary sizes of the respective machine components appear feasible, the next step is to attempt to refine these preliminary sizes using more precise endurance limit information based on a search of the fatigue literature to find (if possible) actual test data for endurance limits pertaining both to the material and its processing selected for the given machine component and to the fatigue test machine whose imposed loading is as similar as possible to the type of loading imposed in service.

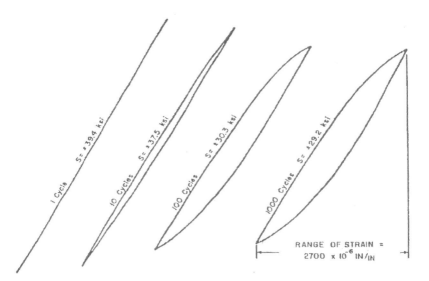

Figure RRM5. Stress-strain hysteresis loops during a strain-controlled axial-load test employing a mild steel specimen. In this extreme example, the first cycle is actually a hysteresis loop with an unloading stress-strain trace that, when plotted, appears to be the same as the stress-strain loading trace. Nevertheless, we know that, because the semi-range of strain, 1350 micro-inches, is very close to the strain at yielding in a conventional tension test, considerable irreversible micro-plastic behavior must have occurred during the loading of this specimen. As loading and unloading continued, this micro-plastic strain accumulated and generated a hysteresis loop that *broadened* with a concurrent *decrease* in the maximum value for the axial normal stress. Metals with this behavior are said to be cyclic softening. In turn, *cyclic softening* behavior necessitates the use of the associated *cyclic yield strength* in fatigue design, rather than the conventional static yield strength. It also raises the issue whether a strain-controlled fatigue test is relevant in machine design given any cyclic softening metal.

Remark: See my topic "Strain-Controlled (Low-Cycle) Axial-Load Fatigue Tests".

CHAPTER 5

Fatigue Crack Initiation and Opening Mode Propagation

F**ATIGUE** cracks are initiated in metals by alternating (cyclic) shear stress and propagated by alternating (cyclic) normal stress. Fatigue crack initiation is typically explained in terms of alternating (cyclic) shear stress causing slip on a specific plane in one direction and then in the opposite direction. But eventually this slip becomes constrained and further slip on this plane ceases. Then cross-slip occurs and moves to an adjacent plane. (Recall the numerous parallel slip planes in the tension test on a HCP whisker.) The alternating (cyclic) stress causes slip and cross-slip back and forth on adjacent planes until intrusions and extrusions occur. The first such observation of these intrusions and extrusions was by Ewing and Humfrey in 1903. See Figure CI1. Intrusions and extrusions are illustrated schematically in Figure CI2.

When in the fatigue process nature decides the initiation process is complete, the fatigue process then continues by crack propagation. (I do not know how nature determines when this transition begins.)

Opening mode crack propagation is illustrated in Figure CP1. In (b) the top and bottom crack surfaces start to move apart due to the increasing axial tensile normal stress imposed on the specimen. In (c) this movement causes massive yielding at 45 degrees to the respective crack surfaces at the extreme end (tip) of the crack. In (d) the crack tip *blunts* under even higher axial tensile normal stress resulting in a crack tip geometry whose length along its blunted surface is much longer than the blunt length. When the imposed axial normal stress changes from tension to compression, the crack attempts to close and the blunted crack tip is force to collapse, causing a small *forward* movement of the resulting compressed (collapsed) crack front. The forward movement of the crack front caused by each cycle of crack tip blunting and collapse in the opening mode fatigue crack growth (propagation) process is appar-

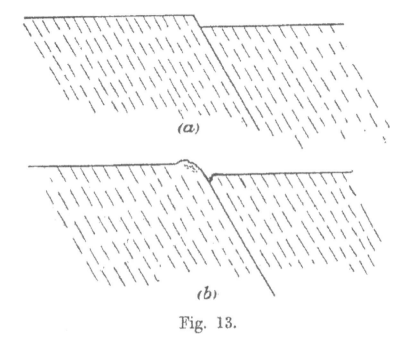

Figure CI1. Ewing and Humfrey Figure 13 sketch of (a) unidirectional slip as in a tension test. (b) alternating (cyclic) slip as in a rotating bending fatigue test. Ewing and Humfrey used a metallurgical microscope to observe formation of extrusions and intrusions on the polished surface of a cantilever fatigue specimen under rotation bending.

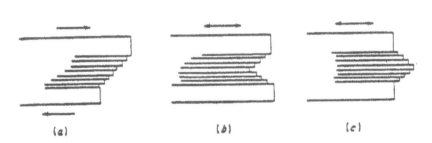

Figure CI2. Wood's schematic model for the creations of intrusions and extrusions by reversed slip and cross-slip. (a) unidirectional slip and cross-slip, (b) and (c) reversed slip and cross-slip. Wood cut sections at a slight angle to the surface of a fatigue specimen to exaggerate the size (depth and length) of intrusions and extrusions.

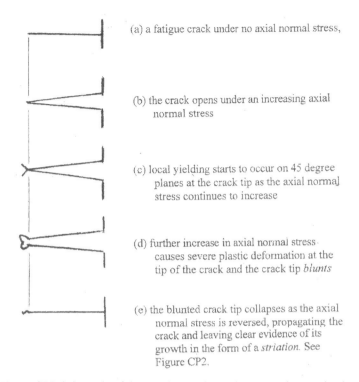

(a) a fatigue crack under no axial normal stress,

(b) the crack opens under an increasing axial normal stress

(c) local yielding starts to occur on 45 degree planes at the crack tip as the axial normal stress continues to increase

(d) further increase in axial normal stress causes severe plastic deformation at the tip of the crack and the crack tip *blunts*

(e) the blunted crack tip collapses as the axial normal stress is reversed, propagating the crack and leaving clear evidence of its growth in the form of a *striation*. See Figure CP2.

Figure CP1. Schematic of the opening mode crack propagation mechanism.

ent when the undamaged failure surface is examined using a scanning electron microscope. See Figure CP2. The fatigue *striations* represent the respective locations of the crack front as it propagates through the metal, cycle by cycle. This crack growth process continues until the fatigue crack reaches a critical length at which time abrupt fracture take place.

Remark: Striations as distinct as in Figure CP2 were obscured by complex microstructures. For example, I did not see striations in mild steels at the magnifications that I used for my SEM images (3000X) in the early 1970's. Striations as distinct as in this figure were obtained only for fatigue cracks in solid solution metals with relatively large grains.

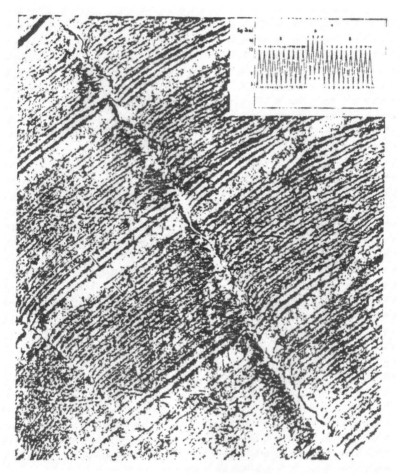

Figure CP2. Scanning electron mircoscope image of a fatigue crack propagating under programmed opening mode normal loading. The number of striations exactly matches the corresponding numbers of normal stress cycles imposed in the programmed loading.

CHAPTER 6

Scatter of Replicate s_a-N Curves and Endurance Limit Datum Values

FIGURE 1(a) displays a typical faired s_a-N curve and the scatter of its datum values pertaining to one of numerous nominally identical *special* rotating-bending tests conducted by ALCOA. Each of these tests were special because all its test specimens were machined from the same World War Two propeller blade. (Propeller blades were generally regarded as having the highest uniformity and the best quality of any aluminum product ever produced by ALCOA). In turn, Figure l(b) displays several *faired* s_a-N curves, each pertaining to a *different* propeller blade that was *nominally identical* to the propeller blade whose faired s_a-N appears in Figure l(a). Note that the differences between these several faired s_a-N curves dwarfs the scatter associated with the (relatively minor) deviations of the individual datum values from their associated faired s_a-N curve. Unfortunately behavior is (almost) universal in laboratory s_a-N data, viz., the *within* variability is (almost) always negligible compared to the *between* variability.

Figure 2 provides another example of this universal behavior. It displays s_a-N data by Nishijima that pertains to 11 melts of two plain carbon steels. Only the highest and lowest s-N curves are plotted for their respective melts in this diagram, along with their associated datum values identified by shading. The unshaded datum values pertain to the remaining intermediate nine melts whose s_a-N curves are not plotted or identified. Again, the scatter of the respective datum values, irrespective of the given melt, are negligible relative to the differences between the faired s_a-N curves pertaining to the respective melts.

I believe these two examples suffice to illustrate the problem with establishing a reasonably precise estimate of even the median endurance limit of mild steels. The statistical distribution of endurance limits is, in one perspective, not the distribution of actual fatigue data, but rather the distribution of what I generically term *batch-to-batch* effects. This creates a serious and costly problem: (i) It casts suspicion on the veracity of virtually all published fatigue data. (ii) It requires that a design analysis methodology that masks these batch-to-batch effects by adopting overly conservative endurance limits estimates and/or larger naive factors of safety. (iii) It absolutely requires appropriate *product testing* in well-simulated service conditions.

I will subsequently illustrate how to use mean stress diagrams for specific modes of fatigue loading to estimate the associated endurance limits of mild steels for the purposes of performing design sizing analyses. *I trust that you will understand that such estimates should never be regarded as being precise.*

Remark: You will likely encounter situations in industry that involve design sizing analyses that are allegedly tried and proven. I hope that you will make a serious effort to understand these analyses and why (or how) each is allegedly tried and proven.

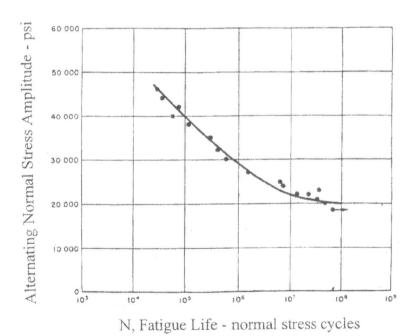

Figure 1(a). A typical faired rotating-bending s_a-N curve and its associated datum values for specimens machined from a World War Two aluminum propeller blade.

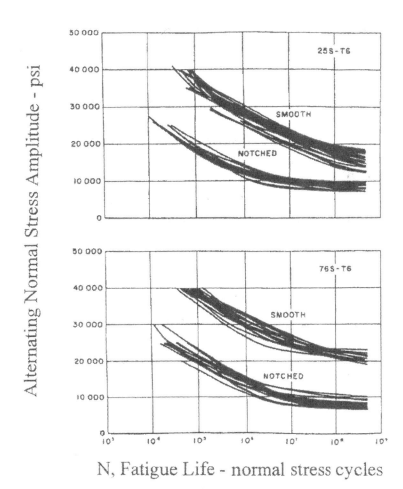

Figure l(b). Several faired rotating-bending s_a-N curves for nominally identical World War Two propeller blades.

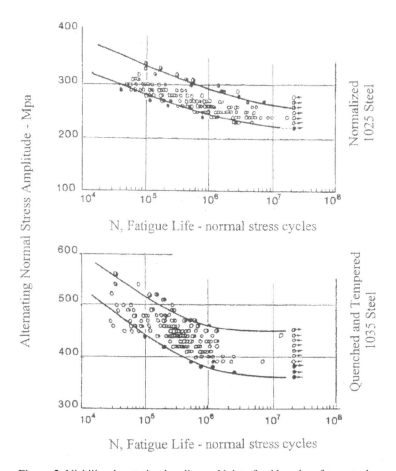

Figure 2. Nishijima's rotating bending s_a-N data for 11 melts of two steels.

Remark One: These two examples suffice to illustrate the problem with attempting to estimate a median value for the endurance limit (fatigue limits at long lives) of mild steels (of aluminums). The distribution of endurance limits (fatigue limits at long lives) is, in one perspective, not the distribution of fatigue data, but rather the distribution of what I call *batch-to-batch effects*.

Remark Two: Note the magnitude of the range of (estimated) values for the (median) endurance limit pertaining to only 11 faired s-N curves. (How would you estimate the magnitude of the range of (estimated) values for the (median) endurance limit pertaining thousands of melts?)

CHAPTER 7

Service Fatigue Failure Exhibits

THE three most common cause of machine component failures in service are fatigue, wear, and corrosion. Wear and especially corrosion are typically visible to the naked eye well before either approaches its respective failure state. Fatigue is much more insidious in that (i) it is usually difficult to detect visually, and (ii) failure occurs abruptly without warning. Fatigue is by far the dominant mode of failure for machine components.

Laboratory fatigue tests generate fatigue failures that seldom look like service fatigue failures. The primary reason is that laboratory fatigue tests impose alternating stresses that have a constant amplitude. Accordingly, laboratory failure surfaces are typically smooth and relatively flat, exhibiting few, if any, distinctive features.

Service failures, on the other hand, have very distinctive failure surfaces associated with their specific service load-time history. Typically the fatigue crack initiation sites are evident and the resulting common fatigue crack front(s) display numerous distinct regions of much faster and much slower fatigue crack growth due to relatively abrupt increases and decreases in service stresses imposed for different durations. This stress history generates a number of what I call *stop marks,* where the rate of growth of the crack front abruptly changes.

At one time I had several hundred fatigue failure exhibits that covered almost the entire range of machinery components and their materials. But time moves on and so my collection and my fatigue machines were lost during a lab renovation. I knew I could not save more than a very few small failure exhibits so I hastily selected some of my fatigue favorites. I do not assert that these exhibits have any coherent relationship. But I believe that a few of these exhibits may serve to illustrate certain fatigue concepts.

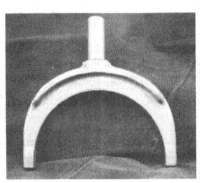
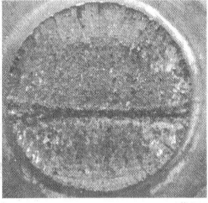

Figure FE1. Motorcycle Transmission Shift Fork. The fatigue failure occurred at the sharp stem-fork transition. This failure is interesting in that although it is due to reversed (two-way) bending, the two sides differ markedly. The longer fatigue crack surface is remarkable in that it has no obvious stop marks visible by the naked eye and looks much like a conventional laboratory test fatigue failure. The failure surface exhibits marked dishing due to the extremely large stress concentration associated with the sharp stem-fork transition, but recovers its elevation rapidly as it grows. The shorter fatigue crack surface exhibits even more dishing and has three distinct early stop marks. The elevation of its crack front is well below that of the longer fatigue crack front when these two fronts approached each other. Abrupt fracture occurred along a thin more or less horizontal strip across the entire diameter. The width of abrupt failure clearly highlights the difference between static and fatigue strengths. Note that each fatigue crack initiated at numerous locations and slightly different elevations before forming a single crack front. The numerous short radial lines are evidence of this behavior. Overall, this failure surface indicates a very high stress concentration with a remarkably uniform bending stress amplitude associated with shifting.

Figure FE2. Another failure exhibit that clearly highlights the difference between static and fatigue strengths. This exhibit is an old time automobile fan with its blades riveted to the its spider. Note (i) the two higher elevations where the fatigue cracks initiate on the shaft fillet, (ii) the several faint concentric stop marks, and (iii) the very small (eccentric) abrupt fracture.

Figure FE3. Another reversed (two-way) bending fatigue failure, but with a more complicated load-time history. This exhibit is a nine inch long, one inch diameter zinc-plated machine screw. The fatigue failure occurred at the sharp transition from the shank to the hex head portion of the screw. This failure surface exhibits short radial lines, top and bottom, similar to those in Figure FE1. After a while the respective fatigue crack growth inward suddenly stopped, creating very distinct *stop marks,* top and bottom. Then, although the service alternating stress was so small that respective fatigue cracks had stopped propagating, the two sides of each crack continued to rub together and oxidize. This action produced two regions of very smooth almost shiny surfaces. (The lighting falsely causes the two different appearances in this photo.) Then the fatigue cracks on each side display a region of very fast crack propagation, followed by another region of very slow crack propagation. Eventually both crack fronts propagated to an abrupt fracture which has a bipedal appearance caused by one side sloping downward and the other sloping upward somewhat akin to shear lips in reversed directions in a cup and cone tension test specimen failure.

Remark: Most people think that because you tighten a machine screw when assembling components, the machine screw experiences only, or primarily, tension in service. Not true. In fact, you can quote me as follows: "If you really want to know what kind of loads are *actually* acting on a given component or member, wait until it fails and then examine the failure surface."

Figure FE4. An example of a fretting fatigue failure. This tapered spindle experiences the equivalent of a press-fit in a hub. Fretting initiated fatigue cracks at two locations on each side of the reversed (two-way) asymmetric bending failure. The larger of the two sets of fatigue cracks had both of its cracks propagate slowly while their surfaces rubbed and oxidized creating very smooth dark surfaces. Then both fatigue cracks propagated much faster exhibiting numerous faint stop marks until they combined to form a single crack front near the spindle center. This crack front then propagated very rapidly to failure. Its start is well-defined by a very pronounced stop mark. The smaller of the two sets of fatigue cracks also propagated slowly while their surfaces rubbed together and oxidized creating a similar, but much smaller set of very smooth dark surfaces. These surfaces had similar elevations, so their respective crack fronts quickly combined to form a single fatigue crack front. This crack front then propagated slowly exhibiting very faint stop marks along its growth. Finally this fatigue crack propagated rapidly while simultaneously plunging in elevation until abrupt fracture took place along a narrow strip that is much darker in appearance in this photograph.

Figure FE5. Torsional fatigue failure of a splined shaft that rotates the tank of a cement mixer truck. The cement slurry is constantly being lifted by interior baffles as the tank rotates, but falls off these baffles after further rotation of the tank. This mixing process causes a continual fluctuation in the torque required to rotate the tank, leading to fatigue failure of its splined drive shaft. I have seen many failures of this type and almost all are what I call *siblings*, because of the nominally identical appearances of their failure surfaces. (When *sibling* failures occur, it is clearly a serious problem.)

Service Fatigue Failure Exhibits 53

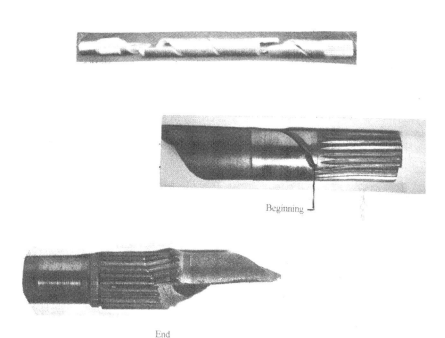

Figure FE6. A torsional fatigue failure of a hollow *hardened* steel automobile transmission shaft approximately ten and one-half inches long. The extraordinary abrupt fracture was caused by small fatigue crack that initiated at the runout of a single spline at the right hand end of the shaft. Then a *square* fracture crack ran along that spline, at *ninety degrees* to the shaft surface, to the right hand end of the shaft. At the same time, a *forty-five degree slant* fracture crack ran along the shaft in a forty-five degree spiral until it ran into the spline at the left hand end of the shaft. Finally, a square fracture ran longitudinally between the spiral fracture cracks at each end of the shaft.

Remark: The fundamental issue associated with this fatigue failure example is that a very small fatigue crack can cause abrupt fracture of a hardened steel component.

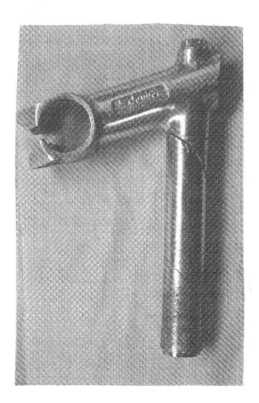

Figure FE7. A torsional fatigue failure of a bicycle handle-bar adjustment clamp. The hollow stem has a saw-cut at the bottom so that a long machine screw can be tightened in its tapered nut to expand the stem inside of the tubular portion of the front wheel fork, to adjust the height and rotational orientation of the handle bars. Service operation caused a fatigue crack to initiate at the top of the saw-cut where the stem experiences only torsional loading. The fatigue crack propagated in a spiral at forty-five degrees up and around stem wall until it reached the yoke. It then propagated around the yoke bottom and then resumed growing at forty-five degrees *downward*, until it ran into the spiral crack that had previously grown *upward*, at which time it could not propagate further.

Remark: The "termination" of this fatigue crack reminds me of an old time method of attempting to stop the further propagation of a fatigue crack. A hole was drilled at the tip of the crack, drastically reducing its stress concentration, thereby sometimes preventing further crack propagation, and sometimes not.

Figure FE8. Fatigue failure of a very large eyebolt with a welded nut to prevent it from loosening. Although a fatigue crack typically initiates only at the first thread, a fatigue crack also initiated at the second thread directly opposite to the initiation site for the first thread in this exhibit. Both cracks propagated for a while and then underwent a pronounced dwell period when their respective fatigue crack surfaces rubbed one another to become smooth and oxidized. Prior to this dwell period, the longer fatigue crack exhibited numerous small radial cracks before converging to a single crack front. This crack front then exhibited numerous stop marks and one-dimensional dishing, continually dropping in elevation. After the dwell period the fatigue crack front continued to drop in elevation for a while and then leveled out and started to increase slightly in the vicinity of the last two very distinct stop marks. Abrupt fracture occurred vertically upward in the region of the second thread.

Remark: Fatigue cracks not only propagate by normal stress, fatigue cracks propagate *perpendicular* to the direction of the normal stress. This is the reason that dishing is often observed in fatigue failures, and the change in fatigue crack surface elevation can be quite pronounced (as in this fatigue failure example).

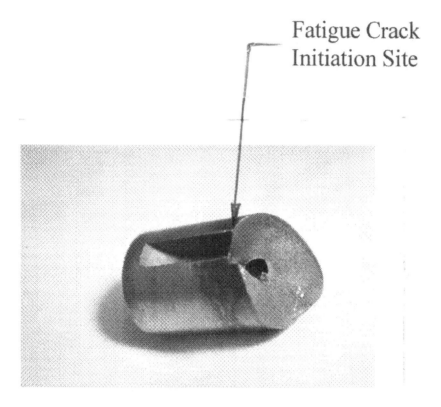

Figure FE9. Fatigue failure of a shaft with a Woodruff keyway. Note that fatigue failure initiated at one runout corner of the keyway at the shaft surface, not at its center where it is deepest. Shafts with sledrunner keyways also fail in fatigue at one of their runout corners regardless of location of the actual end of its key. Shafts with profile keyways fail in fatigue at the intersection of the shaft surface and the edge of one side or the other of the key seat at the beginning of its end radius regardless of the actual location of the end of the key.

Remark: Published values for the endurance limit reduction factor pertaining to sledrunner and profile keyways were obtained by laboratory tests on specimens without keys in the keyseats.

CHAPTER 8

Miscellaneous Failure Exhibits

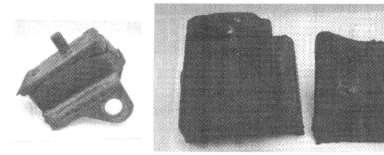

Figure MFl. Fatigue failure of an engine mount. The fatigue crack propagated through this rubber compound with an opening mode mechanism analogous to that in metals. Thus this failure surface exhibits numerous stop marks and various elevations.

Figure MF2. Static yielding of a small square key. This *sliding* form of deformation is quite rare, and even more so, because the service overload did not cause the key to separate into two pieces.

Figure MF3. A *typical* torsional failure of a straight rod with a hexagonal cross-section that drives the oil-pump in an automobile engine. When the oil pump seizes this drive shaft *twists several full revolutions* before fracturing.

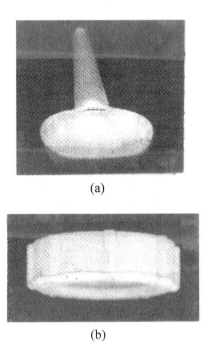

(a)

(b)

Figure MF4. Stress-corrosion cracking failures. (a) This stamped aluminum measuring spoon had massive plastic deformation during its forming operation at the junction of the handle and the spoon that produced a tensile residual stress state. In turn, this residual stress acted in conjunction with moisture to form a stress corrosion crack. If the tensile stress is large enough or if the corrosive environment is strong enough, stress corrosion cracking is transgranular, otherwise it is intergranular. When the British fought in India, stress corrosion cracks in brass cartridges were called "season cracking" because it was much more prevalent during the rainy season. (b) Some plastic parts are also susceptible to stress corrosion cracking. Over-tightening gas and oil caps on jet-skis and on the threaded fittings on sink traps provide the steady tensile stress required to cause stress corrosion cracking.

Figure MF5. (a) Static torsional overload failures of mild steel shafts with a spline that has an undercut at its end. These static overload failures are typically called "ductile failures" because they occur at ninety degrees to the longitudinal axis of the shaft, as opposed to what is typically called "brittle failures" that occur at forty-five degrees to the longitudinal axis of the shaft. (b) A "ductile torsional failure" of a spline without an undercut. It is common for a spline to fail where the mating spline ends. (c) A "ductile torsional failure" of the spline on a shaft where the mating internal spline of a spiral bevel gear ends.

Caveat: Be careful of people who use the words "ductile" and "brittle", especially if they have a mechanics background. Ductile should connote a growth and coalescence of voids crack growth mechanism in mild steels and brittle should connote a cleavage mechanism of crack growth (usually ending in abrupt failure).

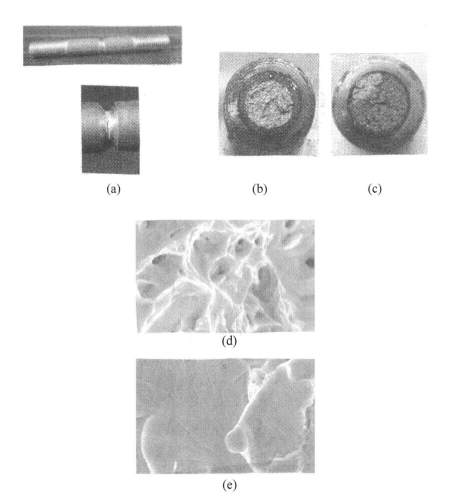

Figure MF6. (a) A sharp V-notched tension specimen of normalized AISI 1020 steel, pulled in static tension to demonstrate that the *crack initiates at the notch root and propagates inward.* Fracture appearances (b) with really good alignment (c) with really poor alignment. Nevertheless the test procedure for both (b) and (c) is the same and the respective outcomes cannot be predicted in advance. The alignment of tension specimens with threaded grips is highly variable, especially if threads are chased, but is only somewhat better even with ground threads. The crack that initiates at the notch root is "ductile", viz., its initiation and propagation mechanism is growth and coalescence of voids, whereas the subsequent abrupt failure (fracture) is "brittle", viz., its propagation mechanism is transgranular cleavage. (d) A scanning electron microscope micrograph of crack propagation by growth and coalescence of voids in "ductile" region of (b). In turn, (e) a scanning electron microscope micrograph of the "brittle" transgranular cleavage crack propagation region culminating in abrupt fracture. Although the boundary between these two failure surface regions may appear to be distinct in photographs (b) and (c), there is actually a very narrow region consisting of both "ductile" and "brittle" modes of failure.

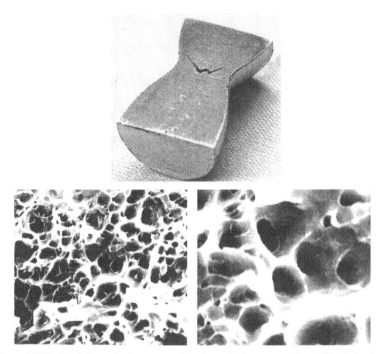

Figure MF7. (Top) A vertical cross-section through the center of a 1100-F aluminum tension specimen pulled past necking and *almost* to fracture, demonstrating that *the crack initiates at the center of the specimen and grows outward.* (If you do not wish to gamble on finding an internal crack by sectioning the specimen, you can X-ray the specimen beforehand.) (Bottom) I regret that I cannot get good Xerox copies of my SEM micrographs for 1100 aluminum taken at 1000X and 3000X that demonstrate the growth and coalescence of voids. Hopefully the following Xerox images provide some perspective.

Remark: This demonstration is only possible using a few select materials which exhibit a very large reduction in area in a tension test. I do not know of anyone who has been able to demonstrate this internal crack in AISI 1020 normalized steel or any other mild steel. Nevertheless, given broken tension specimens of AISI 1020 normalized steel that display the classical cup-and-cone failure, both the cup and the cone exhibit growth and coalescence of voids. The cup portion is flat, oriented at ninety degrees to the longitudinal axis, almost perfectly round and concentric, and it grows to final rupture when the ductile crack growth abruptly changes direction to forty-five degrees from the longitudinal axis thereby forming the concentric shear lip which is referred to as the cone. The only difference between the cup and cone when viewed using a scanning electron microscope (SEM) is the voids along the shear lip are stretched in the direction of sliding.

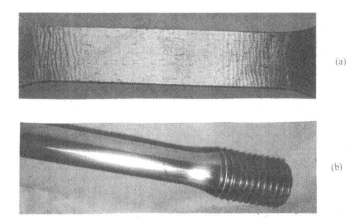

Figure MF8. Luder's bands. Yielding occurs abruptly in a tension test employing AISI 1020 normalized steel specimens, when a wedge of plastically deformed material called a Luder's band forms contiguously to its adjacent elastic material. Each Luder's band exhibits uniform plastic strain that is approximately equal to the strain at the yield point elongation. (a) A dogbone tension test specimen with its *entire* gauge length consisting of Luder's bands. Then the plastic strain along the entire gauge length is equal to the yield point elongation. (b) A conventional (polished) round 0.505 inch diameter cylindrical tension specimen exhibiting a *single* Luder's band. It forms an elliptical trace on the cylindrical surface because it is oriented at forty-five degrees to the longitudinal axis of the tension specimen. In contrast the Luder's bands in (a) are oriented at forty five degrees to the dogbone specimen surface.

Remark: When mild steel bars with a rectangular cross-section are bent with the tension side up, Luder's band that are straight across the top are oriented at forty-five degrees on the front surface; whereas Luder's band that are straight across the front surface are oriented at forty-five degrees on the top surface.

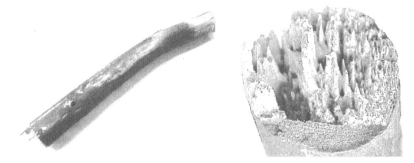

Figure MF9. Fatigue of a wood handle of a shovel. The fatigue crack started on the left-hand side of the shovel indicating that I am a left-foot stomper.

Miscellaneous Failure Exhibits **63**

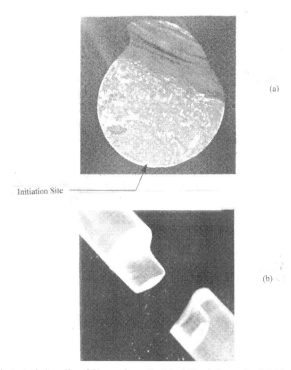

Figure MF10. A static bending failure of a cylindrical Plexiglas rod exhibiting the typical bending failure profile with its distinctive tail. The tensile crack initiated at the bright spot in (a) and propagated rapidly past the neutral axis well into the compressive region before abruptly turning ninety degrees to become perpendicular to the tensile normal strain due to Poisson's ratio. The surface of this tail exhibits Wallner (stress wave) lines. Sometimes this bending failure terminates in a double-tail which "explodes" downward.

Figure MF11. A die-cast bracket exhibiting extreme porosity. Internal defects like this kind cannot be detected by visual inspection. This type of defect is typically attributed to a plugged vent which prevent air from exiting the die.

Remark: One of the more common causes of defects that are not visible by inspection are due to cores shifting their positions during casting.

Figure MF12. An internal defect called *central bursting*. It is caused by excessive deformation during extrusion. This automotive rear axle was known to be susceptible to central bursting so that all axles were ultrasonically inspected. Nevertheless this axle passed inspection but failed when quenched following induction hardening.

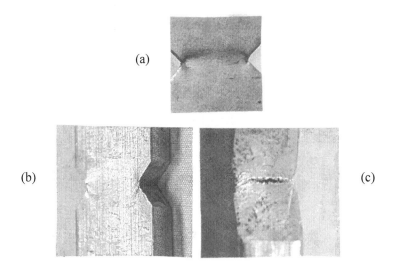

Figure MF13. Double-edge-notched tension test failures for thin and thicker aluminum scrap materials. The thin specimen was sheared from trimmings found in a scrap barrel. I found the thicker specimen on the shop floor. I then used the ninety degree notch shear to notch each specimen. (a) The thin specimen exhibits extreme localized dimpling across the specimen except at the locality of the roots of the notches. The subsequent *v-slant* failure across the central cross-section is actually two shear lips opposed to one-another. The thicker specimen also exhibits dimpling, but not as extreme or as localized as for the thin specimen. There was, however, much more extreme constraint at the roots of the two of these edge notches. (c) When looking directly at the root of each of these notches you will see a crack initiate at the specimen center and tunnel into the interior of the specimen. The subsequent failure *in cross-section* resembles a cup-and-cone failure, viz., it has a flat central crack (formed by growth and coalescence of voids, commonly termed a *plane-strain* failure; and the respective shear lips (also formed by growth and coalescence of voids), commonly termed a *plane-stress* failure.

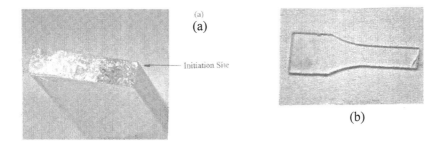

Figure MF14. Dogbone Plexiglas tension test specimens exhibit abrupt tensile test failures that always originate at one of the routed edges, typically at one of its corners. The crack origin is easily recognized because the initial crack growth is slow, creating a small smooth surface that readily reflects light as in (a). Once this original crack reaches a *critical length*, abrupt fracture commences and the stored strain energy in the tension specimen is instantaneously starts to dissipate as this initial crack grows and exhibits numerous "micro-cracks" resembling *scales* that makes the crack surface to appear to be opaque. Although this initial crack grows very rapidly, it cannot exceed the speed of sound in this material. Accordingly, the initial main crack may have to branch into multiple smaller cracks to dissipate the stored strain energy, (b). In turn, these smaller branches also exhibit numerous even smaller "micro-cracks" along their respective surfaces, thereby making a large portion of the failure surface appear even more opaque.

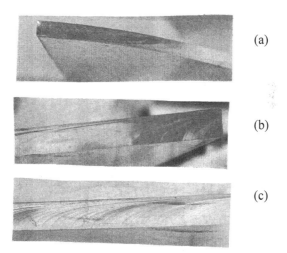

Figure MF15. In glass and most thermoset plastics, the fracture surface will display parabolic stress wave traces as the crack propagates. These parabolic traces are symmetrical about the center of the material thickness when the loading is tensile and open toward the *direction* of crack propagation. However, for bending fractures, only the tensile stress portion the parabolic traces actually form and these traces open toward the tension side of the bending stress. Note that the fracture origin in (a) is easily recognized. The origin in (b) is found by back-tracking the parabolic traces. (c) Stress wave traces produced by a hockey puck hitting a Plexiglas screen

CHAPTER 9

Miscellaneous SEM Micrographs

THE metallurgy department in Ann Arbor bought a scanning electron microscope (SEM) around 1970 or 1971. Fortunately I was able to use this SEM to examine various failure surfaces. I ran laboratory tests such that I knew the actual loading and the associated mode of failure. Then I would use the SEM to examine these failure surfaces at what was then an *astonishing* magnification.

(a)

Figure MSEM1. I ran rotating bending fatigue tests using a cantilever-type machine with ground one-half inch diameter air-hardening (alloy steel) drill rod specimens, notched by milling either a profile or a sled runner keyway at its longitudinal center. My test machine had one-half inch collet grips so these specimens required no additional machining or polishing. In turn, I set the limit switch at a deflection such that it would stop the fatigue test before abrupt failure occurred. Then I pulled the cracked specimen to failure in a tension test machine to prevent failure surface damage. Micrograph (a): The (prior) fatigue crack failure surface. Micrograph (b): The (subsequent) abrupt (brittle) fracture by cleavage. Micrograph (c): The transition from (a) to (b), viz., a very thin strip of (ductile) growth and coalescence of voids that developed at the tip of the fatigue crack, just before abrupt (brittle) cleavage failure occurred Thus, *three distinct modes* of failure appear in this micrograph, all within approximately one-thousandths of an inch. *Figure (b) and (c) continued on next page.*

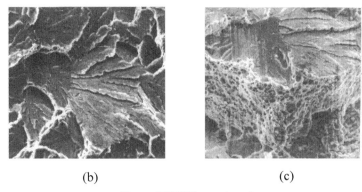

(b)   (c)

Figure MSEM1. *continued*

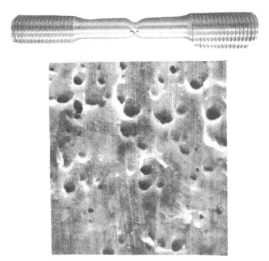

Figure MSEM2. This micrograph demonstrate the sliding mode of failure of 1100F aluminum for the tension specimen with a single three, thirty-seconds of an inch diameter cross-hole. The hole initially becomes elongated as necking occurs and then the sliding failure starts at each side of the hole, growing in a v-shape pointing toward each outer lateral surface of the specimen. In fact, sliding failure terminates with a very sharp (chisel-point) edge on each outer lateral surface. My SEM micrograph pertains to this sliding failure surface and demonstrates that it slices through voids created during the prior elongation and necking portions of the tension test. It is almost like being able to slice through a conventional tension specimen well before the final coalescence portion of growth and coalescence ductile failure takes place.

Remark: Note that the planes of sliding are apparent on the outer lateral surfaces of the tension specimen well before the sliding failure actually reaches these lateral outer surfaces

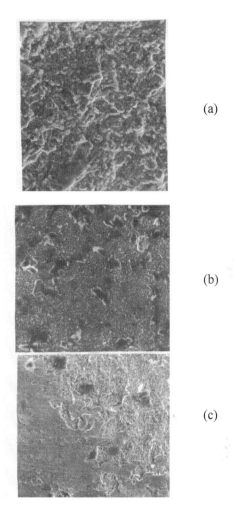

Figure MSEM3. A number of years ago I walked in the ME shop and noticed the tip of a bottoming tap laying on a bench near a vise. Someone had chosen the wrong tap, started it at an angle, and then tried to straighten it, causing the tip of the tap to break off. Evidently the perpetrator has just disappeared. The cutting fluid that he had used was starting to seep on to the fracture surface. So I immediately took it to the metallurgy lab and used alcohol to clean the surface. Then I made the mistake of blowing air on it to dry its surface more rapidly. The air was so moist that I could see corrosion starting to form. I stopped the air and cleaned the exhibit again with alcohol and let it dry in air. Then I took the following SEM micrographs the first chance I got. Micrograph (a) is a region of no corrosion. It is ductile failure by growth and coalescence of voids, but the tap is so hard (about Rockwell C 64) that this may not be readily apparent at 1000X. Micrograph (b) is the "corroded portion" of the failure surface at 250X. Note the *individual (discrete) corrosion platelets.* Micrograph (c) is the scuffed portion tip of the tap exhibit as it rubbed on the remainder of the tap. Note the extreme ductility of this very hard metal surface *when it undergoes constrained deformation.*

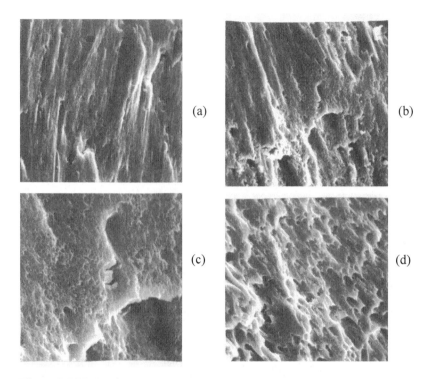

Figure MSEM4. It is common practice to shear sheet metal, plate, and rods to size. The following micrographs pertain to a one-half inch diameter 2024-T351 aluminum alloy rod. Shear failures are actually much more associated with bending than with (textbook) direct shear. Micrograph (a): The beginning portion of the failure surface at 1000X is scuffed by the blade of the shear. Micrograph (b): This bottom portion of the *scuffed* region starts to exhibit (tensile) growth and coalescence of voids at 1000X. Micrograph (c): The same region as Micrograph (b), but at 3000X. Micrograph (d): The tear region at 1000X. It resembles an ordinary shear lip, viz., it exhibits well developed growth and coalescence of voids with the voids elongated in the direction of tearing. Somehow I messed up the direction of tearing *slightly* in (b) and (c), and *drastically* in (d). The actual direction of tearing in (d) is *vertically downward.*

Remark: The rule of thumb for shearing sheet metal is that the beginning third of the thickness is scuffing (smearing) and the final two-thirds is (tensile) tearing.

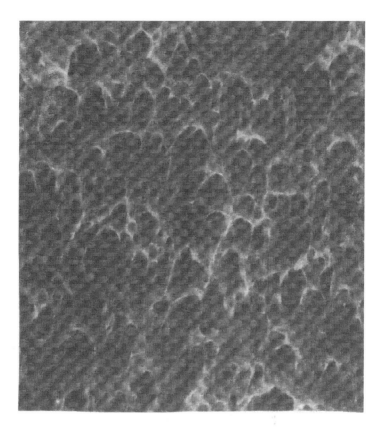

Figure MSEM5. This micrograph is a tensile failure of a conventional 110OF aluminum tension specimen that had undergone a large torsional twist before testing. The resulting tensile failure is obviously not the traditional cup-and-cone failure. Instead, the profile of the resulting necked specimen failure resembles that of a *wolf's ear*. However if I had twisted and re-twisted this specimen the same amount, but in the opposite directions (senses), then this tension specimen would exhibit the traditional cup-and-cone failure.

Remark: This is a powerful example of how prior deformation can establish planes in the material that control the geometry of subsequent failure.

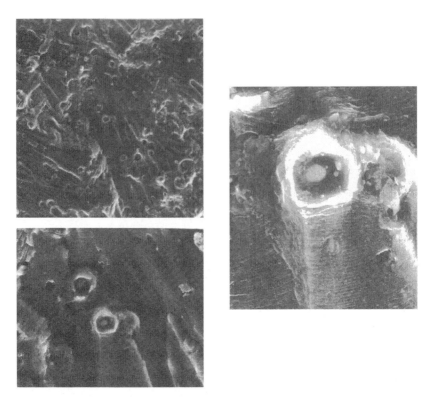

Figure MSEM6. Micrographs intended to illustrate how a precipitate can impede the growth of an opening-mode fatigue crack in a (precipitation-hardened) 2024-T4 aluminum alloy. Consider the central precipitate in the 300X micrograph above, and then observe its appearance in the micrographs at 1000X and 3000X.

Note that when the fatigue crack front meets this exemplar (obstacle) precipitate, it must change its elevation to propagate past this (obstacle) precipitate. Sufficient numbers of such precipitates, of the proper size and spacing, can markedly slow the *fatigue crack growth rate (da/dn)* for a precipitation-hardened aluminum alloy.

Remark: 2024-T4 is an old-time designation for the precipitation-hardened aluminum alloy that is traditionally used for the skin of airframes.

Figure MSEM7. Opening-mode fatigue striations at 1000X in a circumferentially notched one-half inch diameter Plexiglas rod. Note that this fatigue crack growth mechanism is not exclusive to metals.

Remark: Most of my tests on unreinforced plastic material employed a dog-bone specimen one-eighth of an inch thick, with a one-eighth of an inch diameter central hole. It was often possible to observe fatigue striations on broken specimens simply by visual inspection because the striations became very large when one or the other of the crack fronts first reached the immediate vicinity of one edge of the specimen. All you had to do was to view the failure surface at the proper orientation.

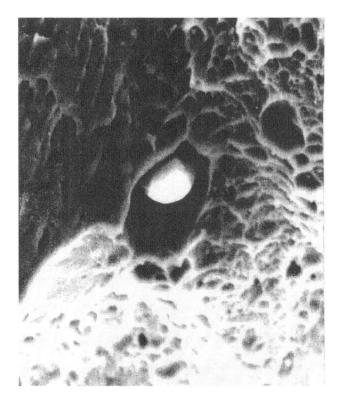

Figure MSEM8. This micrograph is intended merely to present a nonmetallic inclusion located in the midst of a region of ductile growth and coalescence of voids. Typically a microprobe is employed to establish the respective chemical compositions of various inclusions. In turn, it is well known that the number and the chemistry of the various inclusions effectively control the life of rolling-element bearings. Since all bearing manufacturers employ the same 52100 alloy steel, the only difference in quality is the material *"cleanliness"*, viz., the relative absence of nonmetallic inclusions. Cleanliness depends primarily on the number and the effectiveness of the respective vacuum degassing processes. One prominent bearing manufacturer asserts that *each* different chemical composition (type) of inclusion generates a different Weibull life distribution, and that in turn the aggregate life distribution is also a Weibull distribution. The actual fatigue life distribution of a rolling-element bearing is statistically a *competing modes of failure* problem. Moreover, A-basis or B-basis statistical tolerance limits are statistically preferable to B-01 and B-10 lives. But the present methodology is unlikely to change due to the reluctance of bearing manufacturers to release their life data (if any) for competent statistical analysis.

CHAPTER 10

Fatigue Effects

THE *naive* concept of a fatigue effect became widely accepted starting in the early 1930's. The fundamental idea was that it is possible to generate enough fatigue effect data that the endurance limit of any machine component of interest could be estimated. The process for generating individual size effect values would have two stages: (i) fatigue tests employing a *standard test specimen* would establish a reference value, viz., the (median) unnotched endurance limit, and then (ii) given the fatigue effect of interest, say the actual surface finish of the design component of interest, the standard specimen would be appropriately modified and employed in an analogous fatigue test to experimentally establish the value for the fatigue effect as a *proportion* of the (median) unnotched endurance limit reference value. Note a *ratio* is implicit in this definition. Where along the two respective s_a-N curves could such a ratio be credible? Answer: Only where the ratio value is (relatively) constant, viz., at the respective (median) endurance limits for mild steels; or at very long fatigue lives for materials that do not exhibit an endurance limit.

The first problem is to establish the standard test specimen. Unfortunately this has never happened. Rather it has always been a work in progress. Typically the so-called reference test employs an unnotched specimen of more or less a small uniform size and is typically polished using emery cloth in a procedure that has slowly evolved over several decades. However, no specimen polishing technique has ever demonstrated statistical uniformity among fatigue investigators in a round-robin test program. Nevertheless, because both the allegedly standard and its associated fatigue effect tests are conducted by the same fatigue investigator using the same equipment and same material stock, while making all other test conditions as alike as practical, the resulting fatigue effect ratio value is widely accepted as being credible (though perhaps not necessarily either accurate or statistically precise).

The alleged need for generating fatigue effect data provided fatigue researchers an opportunity to publish and thus was greeted with widespread enthusiasm. This created the problem of fatigue researchers looking for a fatigue effect and, wow, finding it.

Size Effect: Size effect ratios are commonly employed in machine design textbook analyses. This use is typically accompanied by some allegedly sensible explanation. However, the authors ignore contrary evidence such as (i) a fatigue investigator found a size effect for his mechanically polished specimens but did not find a size effect for identical specimens that were chemically polished, or (ii) a fatigue investigator who found no size effect between torsion specimens that were three-quarters of an inch in diameter and eleven inches in diameter. (Caveat: It is not easy to fatigue test an eleven inch diameter shaft in torsion.)

The size effect data that I have seen raise two concerns: (i) I am concerned that specimen size and test machine size (capacity) are *confounded.* (ii) I am also concerned whether the original fatigue size effect researchers actually knew the difference between hardness and hardenability. Whenever either a large annealed or normalized cylindrical bar is cut up into slugs to be machined into fatigue specimens, it is very important to understand the initial stock does not have a uniform strength or hardness distribution across its cross section. The static yield strength and tensile ultimate strength, if either were measured, would definitely increase noticeably from the center of the bar to its surface. I know of one fatigue researcher who opted to employ SAE 4340, a deep hardening, sometimes called "through hardening", alloy steel in a size effect test program. He got some lousy datum values and thus went back to examine the microstructures of his specimens and concluded that his original SAE 4340 bar stock exhibited metallurgical banding, a relatively common inhomogeneity problem with this alloy. He then "corrected" his datum values and (wow) he saw a size effect. (I would expect that someone with the knowledge required to correct his datum values would also have known to examine the microstructure of his bar stock before machining his specimens and running his tests.)

I think that much of the so-called size effect, if actually observed in fatigue testing, is actually a static strength effect. The relationship between size and strength at the surface the stock is generally downplayed in stating tensile yield strengths and ultimate tensile strengths for mild steels. The differences have to be really profound, as in cold-drawing or extrusions, before size effect in tensile strengths is actually mentioned in material specifications.

Finally, machine design authors who include a size effect factor in analysis, usually along with a few other effect factors, are essentially employing a multiplicative collection of mini-factors of safety rather than employing a single factor of safety in their analysis. This makes sense only if these mini-factors of safety are actually credible.

Surface Finish: The effect of surface finish may sound like a no-brainier. However, the only surface finish that is actually important is the surface finish at the location of the failure. The prime motivation of a full-scale fatigue test of any type of machine is to learn which component fails first and why that particular component failed. If someone has enough knowledge or experience to state this information before testing, he or she should have conveyed this information to the ignorant designers who foolishly likely over-specified unneeded expensive surface finishes throughout the rest of the machine components.

I have seen the original data underlying some of the design plots giving surface finish effect factors. The supporting data are very limited in quantity and scope and I have serious doubts about its credibility. I regard these curves as mainly conjecture and chutzpah.

Fatigue failures always occur at notches of some kind. Almost all fatigue effect data for notches pertain to the notch as *machined.* A drilled hole is not polished, nor is a sharp groove, a square (snap-ring or o-ring) groove, an undercut, a thread, or a spline, or a keyseat, etc. It is unlikely that even a relatively large fillet is actually polished. Moreover, I have never heard of anyone considering the surface finish for a notch when stating the value of a stress concentration factor. So, like it or not, surface finish is somehow incorporated in the extensive compilation of notched and unnotched endurance limit data that established the empirical values underlying the calculation of the endurance limit reduction factor K_f. (See my topic "Endurance Limit Notch Sensitivity of Round Mild Steel Specimens".) If the surface finish shows strong signs of directionality, the machine marks should be parallel, if practical, to the direction of the normal stress. Also, beware of "invisible" grinding cracks on newly ground surfaces. All machining involves plastic deformation, strain-hardening, and residual stress; but grinding, in particular, can produce both tensile residual stress and grinding cracks if the wrong grinding wheel is used or if inadequate or improper cutting fluid is delivered to the grinding wheel surface. I think that an inadvertent nick, gouge, or scratch more likely to initiate a fatigue crack than surface finish. Moreover, for fun I used to collect fatigue failures caused by stamping an identifying part number at the wrong place on a component.

Shape: There are no fatigue effect values for shape. Rather there are only warnings to avoid sharp corners and feather edges. Nevertheless, I have seen wear create grooves and sharp corners, or even worse a feather edge that resulted in a fatigue failure. Moreover, I have seen die wear change the shape a forged component in a critical location. I must also warn that flanges of sheet metal components can buckle causing *future* failures that do not show up in laboratory tests or in test track endurance tests.

Temperature: There is little if any effect of temperature on the endurance limit up to temperatures where creep and/or relaxation occurs. Then the interaction of creep and/or relaxation with cyclic stress is very complicated and excessive deformation rather than fatigue is generally the dominant mode of failure.

Speed and Wave Form: There is very little effect of speed, if any, on the endurance limit for speeds up to the maximum speeds involved in most machine design applications. The effect of speed if any, is primarily at very slow speeds where the wave form may also involve a *dwell* for some proportion of each cycle. The slowness of the speed can accelerate the growth of the fatigue crack during its crack blunting stage and any corrosion at the crack tip can also accelerate fatigue crack growth. Typically, however, this effect is relatively minor because the vast majority of the fatigue life at long fatigue lives is in the crack initiation stage rather than in the crack propagation stage. Wave form has no effect when the amplitude of the wave form is constant. The only issue with wave form is that it is the original cause of widespread use of hydraulic fatigue test machines (i) to simulate so-called *random fatigue* load-time histories and (ii) to simulate aircraft "missions" and other allegedly relevant stress-time histories. Random fatigue simulations collapsed analytically leaving behind a lot of *very slow* expensive fatigue machines presently conducting constant amplitude tests with simple wave forms.

Machine Effect: By far the most important machine effect on the endurance limit of mild steels is the type of control of the fatigue test machine. All endurance limit data must include a description of its associated test machine. I am absolutely dubious of any fatigue data generated using a crank-driven fatigue test machine. Moreover, my experience indicates that even two nominally identical fatigue machines do (will) not generate "statistically identical" data.

Reliability Factor: This factor could be termed *naïveté on speed.* It is literally insane.

Mean Stress: The effect of mean stress is over-emphasized and poorly handled in the fatigue literature. See my topic, "The Effect of Mean Stress on the Load-Controlled Axial-Load Endurance Limits of Mild Steels". Most published mean stress diagrams are ignorant. It is easy to recognize these diagrams. If the diagram pertains to sheet metal, the test specimen must be restrained from buckling. See my topic "Axial-Load Endurance limits for Mild Sheet and Plate Steels". If the diagram pertains to cylindrical specimens, see my topic, "The Effect of Mean Stress on the Load-Controlled Axial-Load Endurance Limits of Mild Steels". If the notched ultimate tensile strength is equal to the (unnotched) ultimate tensile strength on the mean stress diagram, the diagram is ignorant (because the notched ultimate tensile strength is always much larger than the (unnotched) ultimate tensile strength for mild steel specimens.

Notches: The effect of notches is very complicated. Yet the endurance limit notch sensitivity curves, Figure Q2, in the segment *"Endurance Limit Notch Sensitivity of Round Mild Steel Specimens"* are based on extensive data and thus provide reasonable estimates of notched endurance limits of round specimens. Accordingly, these estimates provide reasonable reference values that form a basis for using unnotched mean stress endurance limit diagrams to estimate the effect of mean stress on the notched endurance limit. These mean stress diagrams in turn provide a basis for computing preliminary component sizes based on naive fatigue factors of safety expressed in terms of plotted distance values pertaining to allegedly relevant load lines *(forthcoming)*.

CHAPTER 11

Cumulative Damage

SUPPOSE that you are just finishing up a s-N test program using a R. R. Moore rotating bending fatigue machine and have observed that the fatigue life data exhibited negligible deviations from the estimated median s_a-N curve. In turn, suppose that you started another test with the next alternating normal stress amplitude of interest and had accumulated a number of cycles that, based on the estimated median s_a-N curve, was 40 percent of the fatigue life to failure at that number of alternating normal stress amplitude cycles. Then you suddenly learned that you had inadvertently selected the wrong alternating normal stress amplitude for the test of interest. What should you do? Well, if the notion underlying cumulative damage were correct, you could change to the correct alternating normal stress amplitude and the resulting fatigue life, based on the estimated median s_a-N curve, would be 60 percent of the fatigue life at this correct alternating normal stress amplitude for a nominally identical specimen without prior testing.

> Stated in more general terms, the proportions of the fatigue life allegedly "used up" during each successive test segment under different alternating normal stress amplitude sum to *one*, regardless of the number of different alternating normal stress amplitude segments employed during the cumulative damage test program, and regardless of their respective magnitudes and the sequence order of these alternating normal stress amplitude.

This cumulative damage concept is *extraordinarily naive*. It should never be taken literally.

When I first joined ASTM Sub-Committee E-9.05 on Aircraft Structures, the airframe testing methodology at that time (1965) was termed block-by-block, viz., a procedure in which consecutive blocks of different constant amplitude loads and associated mean loads were imposed in more or less random order in the block-by-block fatigue test program. This methodology reeked of cumulative damage naiveté. I did not claim that I knew how to run a structural fatigue test, but I understood how complicated this cumulative damage problem was so I always argued their testing techniques were nonsensical. Of course I was continually ignored. Nevertheless, after F-4's, F-5's, F-14's, F15's, F16's and several other aircraft all exhibited fatigue lives of one-eighth to one-tenth of their respective design lives, the United States Air Force finally gave up on the "safe-life" design methodology in 1977 and then opted to adopt the "fail-safe" design methodology. Unfortunately, the technical details associated with the fail-safe design methodology are also equally inept. It requires inspections to find fatigue cracks and then requires associated estimates of the crack growth rates to schedule subsequent inspections. Their ability to find even a well-developed fatigue crack is suspect at best. The calculation of residual fatigue life is perhaps even more suspect. The only advantage the fail-safe methodology is that it hopefully requires more frequent inspections which, if effective, are surely worth the additional cost and effort. But I still have issues with their ability to detect fatigue cracks reliably during their inspections and then to calculate fatigue crack growth rates. And then there is the politics of the FAA.

I got out of ASTM in 1980 because strain-controlled fatigue testing had become dominant. I see little, if any, competent application of such data in machine design.

Remark One: The cumulative damage methodology got its start with an anti-friction bearing manufacturer and it is still found in anti-friction bearing industry catalogs. However, there is no application where the cumulative damage methodology is more nonsensical. The depth of the location of maximum Hertz stress changes as the load changes. Moreover, increasing the bearing load can actually decrease the Hertz stress at the depth location of prior maximum stress, typically resulting in an increased bearing life. The same effect occurs for the Hertz contact stresses in gear teeth.

Remark Two: Nishihara ran tests where he allegedly "used up" one-half of the life in torsion and then tested the same specimen in bending. The result was a longer fatigue life than exhibited by bending specimens without prior testing in torsion. He reversed the process and obtained the same results.

Remark Three: Stulen ran long-life rotating bending tests on titanium specimens and decided to test the specimens to only five times 10^7 cycles because he was worried that he might not finish the entire test program by the time his contract terminated. He then stored these specimens in jars filled with kerosene just in case he might have time later to continue these tests to 10^8 cycles. As it turned out, he did have time to continue these tests. However, he found that the fatigue strength of the stored specimens had apparently increased markedly, viz., the resulting s_a-N curve segment was markedly higher than the previous s_a-N segment. I assert that, even in rotating bending where every surface location nominally experiences the same alternating stress, the specimen will adopt a new set of *different* slip planes when the specimen is removed and then is replaced and the same loading is *re-applied.*

Remark Four: Nishihara ran combined bending and torsion tests where the respective bending and torsional loads were out of phase. The amplitude of the alternating shear stress on the plane of maximum range of shear stress was approximately the same for all phase angles except for 90 degrees, where, strangely enough, *all planes in the specimen surface are planes of maximum range of shear stress.* Then the associated endurance limit, stated in terms of the amplitude of the alternating shear stress, was approximately *twenty-five percent lower.*

CHAPTER 12

The Ugly Truth About Laboratory Fatigue Tests

IF there are any credible laboratory fatigue test data at all relative to machine design, you might think that because of the massive amount of data churned out by the R. R. Moore rotating bending fatigue test machine, it would be the associated endurance limits of unnotched mild steel specimens. But don't believe it. This test machine has never been calibrated while it is running, rather its only calibration is when it is not running. The actual stress-time wave form is a pure cosine curve *with super-imposed higher natural frequencies and white noise.* (We just pretend it is a pure cosign curve.)

Cantilever-type rotating-bending fatigue test machines have the additional problem of attaining proper test specimen concentricity along the specimen from collet to collet (from grip to grip), not just at startup but during the entire test period. The concentricity can change during the testing sufficiently so the resulting vertical component of the vibration of this machine can cause a change in its location unless its base is bolted to the floor. This problem is much more extreme for crank-driven, controlled-deflection, plane-bending machines. The wave form is also a pure cosine curve with super-imposed higher natural frequencies and white noise. But, if the machine base is not bolted to the floor, the vertical component of its vibration at relatively high crank RPM's can instantaneously lift the machine and its base off of the floor. (This vibratory action provides a convenient effortless method to move this machine and its base to another location if so desired.) Finally, axial-load tests are plagued by alignment problems that superimpose bending stresses on the axial stresses. This problem is mitigated somewhat by having a higher mean stress, but pulsating wave forms can cause marked vibration problems unless at least a small minimum load is retained.

No laboratory fatigue test machine is without its specific problems; *but crank-driven displacement-controlled machines are absolutely the worst.* The fundamental truth is that different types of fatigue test machines generate different test results, particularly given a wide range of test loads or displacements. You cannot even be certain the two nominally identical commercially available fatigue test machines will generate results in a paired-comparison test program that fail to reject the null hypothesis that the respective machines generate identical results. Moreover, there never has been a test specimen polishing methodology proposed that has passed a round robin uniformity test. (The idea behind developing a uniform polishing technique originally was to create uniform surfaces for the reference specimens employed in establishing *fatigue effects.*)

The most obvious and troublesome problem with laboratory tests is that the load-time history in laboratory tests typically has little or no relationship to actual service load-time history. Almost all service load-time histories have occasional very large loads interspersed in a long more or less uniform load-time history. The number and intensity of the occasional very large loads that do not immediately cause failure have a pronounced effect on the fatigue life. Many years ago, before United States Air Force gave up on the so-called *safe-life design* methodology, it tried employing various allegedly realistic stress-time histories in fatigue testing the main spars of aircraft wings. But, in developing the C-5A, the contractor could not attain the minimum required fatigue life given the required (contracted) allegedly realistic load-time history. The contractor was forced to get the Defense Department to agree to modify the required (contracted) load-time history by removing the largest of the occasional very large loads. Subsequent fatigue testing with the modified load-time history generated more cracks and exhibited an even *shorter* fatigue life. The Defense Department admitted in 1977 that its *safe life* design methodology was a failure and then adopted the *fail-safe* design (or damage tolerant) design methodology (which is not without its own serious problems).

My next example of problems with laboratory fatigue test machines and fatigue test methodologies pertains to world-class fatigue investigators. In fact three of the fatigue investigators involved were the Chairman of ASTM Committee E-9 on Fatigue at some point during their illustrious careers.

After World War II the United States Air Force convinced the newly renamed War Department (now the so-called Defense Department) to support very extensive fatigue tests on several aircraft materials. The most important series of these tests were conducted by Battelle Memorial Institute in Columbus, Ohio, and by NACA in their Langley Aeronautical Laboratory in Hampton, Virginia. About five tons of each of 24S-T3 and 75S-T6 from two consecutive lots were purchased at one time and employed in the respective tests. (NACA Langley did not even have fatigue testing capability before these tests began.)

Figure UT1 presents the s-N data for 75S-T6 aluminum alloy. I quote from NACA Report 1190, May, 1953: "For the 75S-T6 material the agreement may also be considered very good for medium stresses. For low stresses, the NACA results are lower than the Battelle results, particularly at the stress ratio $R = -1.0$. This tendency was noted early in the tests, when only a small number of tests have been made in either laboratory. In an effort to eliminate the discrepancies, exchange visits of the staffs of

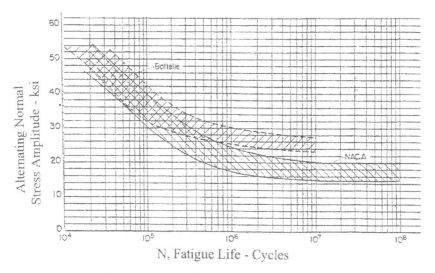

Figure UT1. NACA Report 1190 axial-load s_a-N data for 75S-T6 sheet specimens (with shims to prevent buckling) depicting the unexplained "discrepancies" between NACA and Battelle data at long fatigue lives. NACA later became NASA. Stress ratio R is the ratio of minimum cyclic stress divided by the maximum cyclic stress. Hence, $R = -1.0$ is fully reversed stress or alternating stress.

the laboratories were made, each step in the test procedure was discussed and carefully checked, and additional tests were made in each laboratory. In spite of all efforts, however, it has not been possible so far to reduce the discrepancies further or to explain them."

I have personally asked each of the three fatigue investigators who were formerly the Chairman of ASTM Committee E-9 on Fatigue why these two long life scatter bands did not overlap. None could offer any explanation; they simply did not know.

NACA Report 1190 describes and diagrams the respective laboratory fatigue test machines. Battelle Memorial Institute employed a crank-driven, *strain-controlled* fatigue test machine. Langley Aeronautical Laboratory, which originally did not have fatigue testing capability, built a fatigue machine patterned after an aircraft company design. It was a *load-controlled* machine that operated on a subresonance principle. Its principle is fine, but it is not, in my opinion a well-designed machine. Nevertheless the issue is that *it is incredible that these experts did not employ the same test machines.* Evidently they thought that strain-control and load-control fatigue tests machines generate identical results! What makes it even more discouraging is that a commercially available well-designed subresonance-principle load-controlled fatigue test machine was perfectly suited for these fatigue tests and should have been used in each laboratory. The fatigue area has always been way behind relative to test planning and statistical analyses.

At one time I had two sets of two nominally identical commercial fatigue test machines. After running a number of paired-comparison fatigue-life test programs using the respective test machines as "treatments", I became convinced that the respective machines in both sets were slightly different. Unfortunately there is no way for me to know if either machine gave "correct" data. Nevertheless, I had to decide which of the two test machines that I would use in subsequent fatigue test programs. I chose both.

I know of a situation where during a relatively long test program in the automotive industry that the fatigue testing equipment and test technicians were moved from one building to another building approximately two blocks away. When the fatigue life changed markedly after the move, the issue then was which set of data were actually correct. The *common sense answer:* **neither!**

Hopefully, sooner or later such inconsistencies and the large differences in fatigue life from material batch to material batch will cause a shift to a much wider use of statistically planned comparative fatigue test programs.

The fatigue literature has many comparative tests, viz., the so-called fatigue effect tests that we discuss elsewhere. Unfortunately *fatigue effect* tests are seldom statistically planned. In fact, I recall one comparative fatigue life test program so poorly planned that the investigator actually concluded that *the fatigue life was greater under a severely corrosive environment than in air.* This crazy result was obtained under a contract with the National Science Foundation.

Machine Design Perspective: Each different fatigue test machine has distinctive performance features that subtly modify the resulting value of its endurance limit data, especially (i) what is controlled during the test and how is it controlled, (ii) the rigidity and the construction of the fatigue test machine, (iii) the rigidity and the concentricity or alignment of its fixtures and its grips, and (iv) the specimen type and its polishing. Thus, when searching the fatigue literature for endurance limit data, it is vital to match, as well as possible, the details of the given fatigue test to the details of the given component of interest. Moreover, an endurance limit established by results of a small sample up-and-down test program is much more credible than one established by a s_a-N curve.

Remark: I think that I would be remiss if I did not also tell you that a few years after the *C5A* load spectra fiasco a prominent aircraft structure designer describing his recent tests on another aircraft bragged that he had modified the required (contracted) load-time history by removing the largest of the occasional very large loads to produce *even safer* test outcomes than was required by the contract. He neglected to admit he only learned why this methodology is safer *after* the C5A fiasco.

CHAPTER 13

Ugly s_a-N Curves

FIGURE U1 displays a NASA s_a-N curve that I call ugly for three reasons. (1) It covers so-called fatigue lives so short that fatigue is not the actual mechanism of failure and fatigue lives so long that the s_a-N curve has become almost horizontal. This range of testing defies logic. It is always *good practice* to limit the range of testing, whatever the mode of failure, to the *shortest range of practical interest* in the given design application. This methodology maximizes replication which is essential in subsequent statistical analysis. Most early fatigue investigators conducted s_a-N tests on mild steel that employed only six to eight specimens, with typically all but one or perhaps two at most, tested at stress amplitudes pertaining to relatively short fatigue lives then the only numerical data that they reported was their estimated value for the (median) endurance limit. Surely you would think they would have understood that with only six to eight test specimens, a much better test methodology would have been to test as many specimens as possible at fatigue lives around the *endurance limit*. I have always attempted to gain as much information as possible from my fatigue tests, even to the point of having a microcomputer program to select the stress amplitude for my next test. Test planning before test conducting is always prudent. Virtually none of the tests published in the fatigue literature today actually involves a well-planned test methodology. (2) This sheet specimen will surely buckle at the higher alternating normal stress amplitudes and likely will buckle even at the lower alternating normal stress amplitudes. Accordingly, these specimens had to be restrained from buckling by the use of a lubricated shim on each of its lateral sides. I regard the use of shims as ugly simply because there is no such use of shims in service operation. Accordingly, I also regard the use of the resulting data as irrational and extremely dubious as best. I always prefer a test that simulates as well as practical fatigue loading of the actual component of interest. There are applications where laboratory tests just do not suffice. (3) The range of alternating normal stress amplitudes that I refer to as yielding fatigue involves such large deformations that the term fatigue is actually inappropriate. These deformations are literally insane from a machine design perspective. The range of alternating normal stress amplitudes that I refer to as *deformation fatigue* pertains to deformations that are so large that excessive deformation is very likely to be the actual mode of failure. Most certainly this notched sheet specimen will exhibit observable yielding at the root of the notch if you actually look for it. The magnitude of the observable deformation at the root of the notch will diminish markedly as the alternating normal stress amplitude approaches the nominally elastic range of stress. This sheet specimen does not exhibit a notch strengthening effect so that in this nominally elastic range when the specimen fails, the two sides of the failure surface would fit together almost perfectly to resume the original geometry, if were not for the surface damage occurring as the fatigue test machine slows down and stops. Accordingly, I regard these failures as actual fatigue failures, viz., as legitimate fatigue data.

Remark: Sometimes it is possible to adjust the limit-switch that stops a laboratory test machine to detect fatigue crack growth and stop the machine

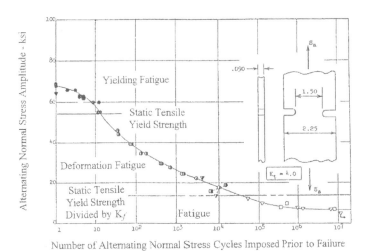

Figure U1. Range of alternating normal stress amplitudes imposed in testing that cause what I refer to as (i) yielding fatigue, (ii) deformation fatigue, and (iii) fatigue. The latter range is the only one of interest in machine design because it contains the portion of the s_a-N curve that is almost horizontal and pertain to nominally elastic behavior, (I shaded the respective published glyphs to emphasize the three respective ranges.)

just before failure would occur. If so, the unbroken specimen (but cracked) can be removed from the machine before it breaks. In turn, if it is pulled to fracture in a tension test, the electron scanning microscope will produce very nice pictures of the fatigue portion of the failure. Otherwise the two sides of the fatigue failure will likely scuff one another and limit the value of the specimen as a fatigue exhibit.

I also recall seeing an ugly s_a-N curve published by an icon in fatigue pertaining to a crank-driven deflection-controlled plane-bending fatigue test with the nominal mean stress so high that its knee was virtually undetectable because the entire s_a-N curve, if plotted, would have been almost horizontal. Moreover, since there were no failures at fatigue lives greater than about 10^4 cycles, it seems that 10^4 cycles would suffice as the estimated location of the knee in the s_a-N curve if it were actually drawn. Any s_a-N fatigue test with imposed strains (stresses) so large as to defy common sense in machine design applications are ugly in my opinion. See my topic "Strain-Controlled (Low-Cycle) Axial Load Fatigue Tests".

Finally, the shape of the s_a-N curve in Figure U2 below is quite unusual. Evidently the aluminum alloy tested had its precipitation hardening process exacerbated by certain accompanying amplitudes of alternating normal stress. Nevertheless the only practical fatigue test data in machine design pertains to the portion of the s_a-N curve is that which is nearly horizontal.

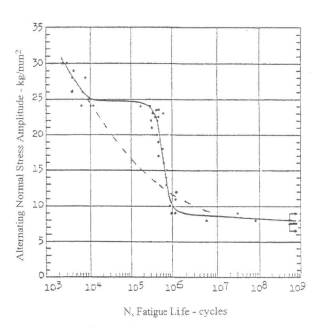

Figure U2. Unusual, and *ugly*.

CHAPTER 14

The Actual Location of the Knee in Rotating Bending s_a-N Curves for Mild Steel Specimens

I present this topic because so many machine design books naively locate the knee in their fictitious s_a-N curves at 10^6 normal stress cycles regardless of the given fatigue effect of interest. In fact, I assert that if you open a machine design book that (i) draws all s_a-N curves with the knee at 10^6 cycles, (ii) draws a Haigh-Soderberg mean stress with the notched and unnotched endurance limit values both equal to the ultimate tensile strength, or (iii) discusses finite life design calculations, you should immediately close that book and open a different one because it is a clear signal that the author has little or no experience in fatigue testing and very likely has little or no actual knowledge regarding competent fatigue analyses.

One of the problems in interpreting published s_a-N data, particularly with regard to establishing a credible endurance limit is that the vast majority of early fatigue investigators were apparently so naive that they often extended the downward slope of the s_a-N curve to some predetermined number of cycles for the knee in their s_a-N curve and its associated endurance limit. About fifty years ago I compiled twenty-two sets of *paired* s_a-N curves with published endurance limit values pertaining to solid and hollow specimens. Each set of these paired s_a-N curves pertained to a different fatigue investigator who tested a same mild steel using the same test machine, the same outside diameter, and the same polishing methodology for the respective solid and hollow specimens. The result for twenty-one of these tests was that the hollow specimen had a lower endurance limit. Yet, the respective differences in these *paired* endurance limits were large enough that twenty-one out of twenty-two having the same type of result was reasonable in my opinion. Rather I think the explanation lies in even a more consistent result, *all* twenty-two had approximately the same downward slopes for the

95

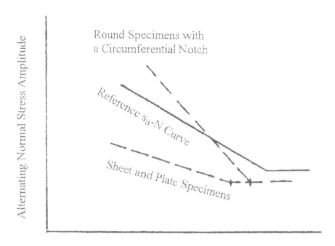

Figure Kl. The Effect of Notches: (a) Notched round specimens exhibit both notch strengthening and constrained yielding at finite fatigue lives. The higher the stress concentration factor the greater the notch strengthening effect and constrained yielding effects, and the greater the fatigue life reduction (retardation) at the knee. (b) Notched sheet specimens exhibit more cyclic softening effects than round specimens, which tends to lower the slope of their s_a-N curve and reduce (retard) the fatigue life at the knee. Both effects are exacerbated by the edge effect associated with the notch.

respective paired s_a-N curves, with the hollow specimen *always* displaying a shorter fatigue life and thus a *lower* fatigue strength along its s_a-N curve. Accordingly, I think that twenty-one of the twenty-two fatigue investigators naively gave too little credence to the alternating stress values for their runout(s) and instead established their respective endurance limits by having the respective knees plotted at the same number of fatigue cycles, viz., they naively and incorrectly gave more credence to their short life data than they did to the long life runout data.

More experienced and wiser investigators like Nishihara have always let their runouts establish their endurance limits and in turn they let the location of the knee be established by the intersection of the horizontal endurance limit and the downward sloping finite life portion of the s_a-N curve.

Remark: It has always amazed me that most early fatigue investigators ran s_a-N tests with about four to six fatigue specimens failing

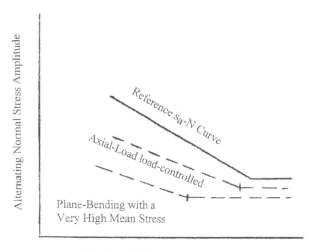

Figure K2. The Effect of Mean Stress: Given axial loading under load-control, the effect mean stress on the s_a-N curve for mild steels is to reduce (retard) the fatigue life at its knee and to lower its downward slope. (The effect of mean stress on the endurance limit is demonstrated in Figure MS2 found in topic "The Effect of Mean Stress on Load-Controlled Axial-Load Endurance Limits of Round Mild Steel Specimens".) These changes are mild compared to strain-controlled plane-bending tests, where a very high mean stress can result in a s_a-N curve that can be so close to horizontal that the location of the knee is difficult to establish precisely. This fatigue life reduction (retardation) is the largest that I have seen for any fatigue effect. I caution you, however, to be leery of any fatigue data pertaining to any crank-driven fatigue machine because strain-controlled fatigue tests suppress yielding and cyclic dependent behavior.

at finite lives and only one or two runouts (at most). Then, often the only numerical data they stated in their published paper was their estimated (median) endurance limit. Thus, it always seemed irrational to me to run so many finite life tests and so few long life tests.

I reviewed a paper manuscript many years ago in which the head of a fatigue laboratory tabulated several sets of s_a-N datum values (without the original faired s_a-N curve) and then had each of his investigators plot the respective s_a-N curves. He then collected the results and superimposed the respective s_a-N plots on the same plot, demonstrating large differences among the individual s_a-N curves. Virtually all of the s_a-N curves in the published fatigue literature are similarly subjective!

You might think that using a least-squares (regression) computer program would solve the problem of subjectivity. It wouldn't. It merely would make the respective plots visually much more consistent, but it would not make them more credible: (1) Least squares (regression) analysis pertains to finite fatigue lives only, not to endurance limits. (2) Least-squares (regression) analysis is typically not valid for materials without an actual endurance limit because the associated finite life datum values typically are *obviously not homoscedastic.* (3) Moreover, *there is no credible methodology to compute a "notch/actor" given unnotched finite life fatigue data.*

In the *comparative* plots in this chapter, I show (i) a typical s_a-N curve pertaining to an unnotched mild steel polished fatigue specimen, and (ii) for *certain fatigue effects,* I show how the fatigue life associated with the knee of the fatigue effect associated s_a-N curve is reduced (retarded) relative to the fatigue life associated with the knee of the given reference s_a-N curve, and whether the associated endurance limit is either reduced or eliminated.

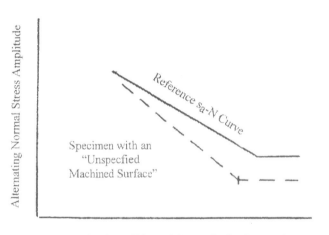

Figure K3. The Effect of Surface Finish: The effect of surface finish on unnotched specimens is more pronounced at long fatigue lives than at shorter fatigue lives. A lathe turned fatigue specimen tested without polishing will exhibit a s_a-N curve with a shorter fatigue life at its knee and a lower endurance limit than a similar polished fatigue specimen. However, unless the surface contains a flaw of some kind, a fatigue crack will almost always initiate at some unpredictable location on the specimen surface. Then the effect of surface roughness is to reduce (retard) the knee and reduce (lower) the endurance limit. However, virtually all fatigue failures occur at machined notches. Then the only surface finish of interest is the surface finish associated with the notch root where the experimental value for K_f also depends on the actual geometry of the notch and the amounts of cold work and residual stress at the root of the notch.

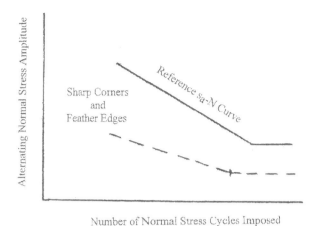

Figure K4. Effect of Shape: A feather edge is absolutely the best fatigue crack originator, followed by a sharp corner edge or deep inadvertent machined groove. Fatigue specimens employed in *fatigue effect* tests may exhibit such shape effects, but well designed machine components do not.

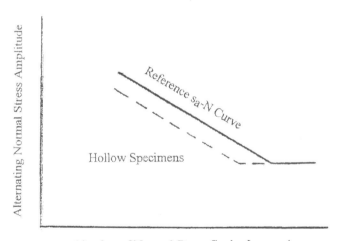

Figure K5. The Effect of Hollow Specimens: I have looked at a lot of hollow specimen s_a-N data and I still do not know the knee is actually affected by this geometry. I think this diagram is most likely to be correct. It simply asserts that the crack, once formed, takes fewer imposed fatigue cycles to transverse a hollow specimen than it does to transverse a solid specimen. Analogously, I think this simple assertion also explains the effect of size on the location of the knee in s_a-N curves for specimens of different diameters. Although I am not convinced that there is an actual size effect for solid specimens in terms of endurance limits (discussed elsewhere), I am confident that larger specimens have their knee at longer fatigue life than smaller specimens merely due to the number of fatigue cycles required to propagate the fatigue crack further.

Remark: I should point out that I also now know of a total 37 paired tests on solid and hollow specimens in two independent fatigue investigations, in which the investigators reported that 34 of the paired tests gave a lower endurance limit for the hollow specimen. Moreover, as nearly as I can discern, these paired tests pertained to the same material, same crank-driven plane-bending test machine, same outside diameter and same surface finish. Although I cannot find my compilation to verify it, I do not think it contained any crank-driven plane-bending fatigue test data. Then, combining my compilation and the crank-driven plane-bending compilation indicates that 55 out of 59 paired fatigue tests gave reported endurance limits lower for hollow specimens. Moreover, I believe that (i) *all* of these tests had their finite life portion of the respective s_a-N curves shorter for the hollow specimens, and (ii) the respective investigators were very likely *biased* by this behavior. I also believe that the respective differences in reported endurance limits are so small that it is extremely unlikely for these reported results could actually be so consistent.

CHAPTER 15

Axial-load Strain-controlled (Low-cycle) Finite-life Fatigue Tests on Round Mild Steel Specimens

AXIAL-LOAD strain-controlled (low-cycle) finite-life fatigue tests have no credible application in either mechanical design or mechanical reliability analyses when many, if not most or all, strain semi-ranges employed in testing are much larger than the strain pertaining to the static tensile yield strength established in a conventional tension test employing a round unnotched specimen. Nevertheless, because such strain-controlled (low-cycle) tests employing round *unnotched* specimens have been so widely conducted, we briefly discuss these tests below.

Originally it was asserted that low-cycle fatigue datum values plotted on $\log(\Delta\varepsilon_{total})$ versus $\log(N_f)$ coordinates followed a straight-line relationship, where $\Delta\varepsilon_{total}$ is the (invariant) range of total strain imposed in each individual fatigue test, as crudely measured by the movement of the upper platen in a lead-screw test machine, and N_f is the resulting fatigue life in cycles. Later, it was asserted that tension test data could be plotted this straight line low-cycle fatigue curve; and that the correct value for N_f was either 1/4 cycle or 1/2 cycle. Eventually it became clear the neither value for N_f was correct and that even the linear relationship was not valid. Nevertheless, the value of 1/2 cycle persisted in a new mechanics model in which it was termed a stress reversal. This new model employed a finite life test methodology in which the respective stress-strain hysteresis loops pertaining to an invariant total range of imposed strain were recorded using a strain-gage extensometer mounted on the (unnotched) test specimen. Then a specific hysteresis loop was chosen for subsequent analysis in which it was asserted that (i) the invariant total range of strain could be partitioned into "elastic" and "plastic" components, and (ii) the associated strain-controlled finite-life fatigue model could be expressed as

$$\frac{\Delta\varepsilon_{total}}{2} = \frac{\Delta\varepsilon_{elastic}}{2} + \frac{\Delta\varepsilon_{plastic}}{2} = \frac{\sigma_f}{E} \cdot (2N_f)^b + \varepsilon_f \cdot (2N_f)^c$$

in which the *actual dependent variable* $2N_f$ is the number of strain reversals (instead of the number of strain *cycles* N_f) and the *actual independent variable* is one-half of the total range of strain $\Delta\varepsilon_{total}$, σ_f is the true stress at fracture established in a conventional tension test, E is the (cyclic invariant) elastic modulus established in this conventional tension test, ε_f is the true strain at fracture established in this conventional tension test, and b and c are constants to be determined experimentally. This mechanics finite-life fatigue model was based on the *absolutely absurd* implication that a tension test is actually a limiting value of a strain-controlled (low-cycle) finite-life fatigue test with N_f equal to 1/2 cycle. Incredibly, it took considerable experimental evidence to convince its proponents that this model was actually untenable. In turn, σ_f simply became σ'_f, the so-called fatigue strength coefficient, and ε_f simply became ε'_f, the so-called fatigue ductility coefficient. Accordingly, the respective "elastic" and "plastic" component fatigue models were subsequently expressed just as ineptly as before, viz.,

$$\frac{\Delta\varepsilon_{elastic}}{2} = \frac{\sigma'_f}{E} \cdot (2N_f)^b$$

and

$$\frac{\Delta\varepsilon_{plastic}}{2} = \varepsilon'_f \cdot (2N_f)^c,$$

whose coefficients were then *naively* estimated by separate (but not independent) "least-squares" analyses. In turn, for each value for $2N_f$ of interest, the ordinate of the (median) est[log($\Delta\varepsilon_{total}/2$)-log($2N_f$)] model was computed as the sum of the ordinates of the (median) est[log($\Delta\varepsilon_{total}/2$)-log($2N_f$)] model and the (median) est[log($\Delta\varepsilon_{plastic}/2$)-log($2N_f$)] model.

Remark One: Recall that my topic "Miscellaneous SEM Micrographs" presented SEM micrographs for understanding the actual physical mechanisms underlying both tensile and fatigue failures. These micrographs clearly highlight *absolute absurdity* underlying this strain-partitioning *mechanics* model.

Remark Two: Because the values of the "elastic" and "plastic" strain components change markedly during the course of each individual strain-controlled (low-cycle) fatigue test (while their sum remains invariant), the respective "elastic" and "plastic" strain components actually reported as datum values pertain to the value for $2N_f$ are established using either the allegedly "stabilized" (no longer changing) strain-stress hysteresis loop or the strain-stress hysteresis loop observed at *one-half* of the N_f value at fatigue failure.

ASTM Committee E-9 on Fatigue conducted a round-robin strain-controlled (low-cycle) fatigue test program in 1980 in which several laboratories each conducted two replicate tests at each of several different values for the imposed total range of strain. Based on the presumptions that strain-controlled (low-cycle) fatigue data are homoscedastic and have $\log_e$-normal fatigue life distributions, Snedecor's central F test statistic was used to test that null hypotheses that the respective *linearized* "elastic" and "plastic" components of the above fatigue model are correct versus the alternative hypotheses that these models are not correct. The *majority* of the laboratories participating in the round-robin submitted data for which the null hypothesis was rejected, given a *Type I* error equal to 0.05, for *each* of the linearized expressions in favor of the alternative hypothesis. (I know, because I analyzed the data.) Thus it has long been clear that this strain-controlled (low-cycle) finite life fatigue model with "elastic" and "plastic" components is statistically dubious. Nevertheless this type of fatigue testing persists even though it imposes ranges of total strain that have no practical machine design application. Even much worse, it is now being presented in machine design textbooks.

Note that to allegedly have practical application in mechanical design, this strain-partitioning model must incorporate a credible way to modify finite-life fatigue data pertaining to unnotched mild steel specimens to account for the effect of the notch of interest. Since an extensometer cannot be placed at the root of a sharp notch, strain-controlled (low-cycle) fatigue tests employing notched specimens are not feasible. Thus some analytical alternative must be found. Enter Neuber's (empirical) rule. However, this rule, if it is ever credible at all in any design application, must pertain only to *plane-stress* applications, and not to *plane-strain* applications. See Figure MS7 in my topic: "The Effect of Mean Stress on the Load-Controlled Axial-Load Endurance Limits of Mild Steel". This work by Ludwig and Scheu in 1923 clearly demonstrated that the sharper the circumferential notch, the greater the notch strengthening (due to the increased constraint at the root of the notch.) See also Figure Q1 in my topic "Endurance Limit Notch Sensitivity of Round Mild Steel Specimens", which demonstrates the notch strengthening effect of a circumferential notch in rotating bending s_a-N data for mild steels. Moreover, axial-load fatigue tests clearly demonstrate that given two round fatigue specimens, one with a fillet and the other with a circumferential groove that have the *same value* for their theoretical stress concentration factors (K_t's); the respective (median) s_a-N curves pertaining to both specimen geometry's are remarkably similar at their endurance limits; but for finite fatigue lives the specimen with a circumferential groove has markedly higher (median) fatigue strength than the specimen with a fillet and this difference *increases* as the finite-life decreases. Moreover, although the endurance limit notch sensitivity q is between zero and one by *definition*, its value can be *negative* if it is inappropriately applied to the finite life portion of the (median) s_a-N curves for the respective specimen geometry's.

APPENDIX

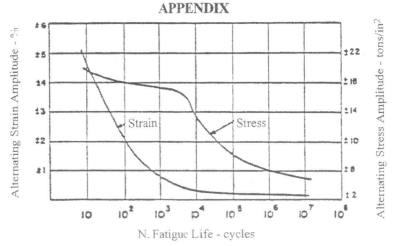

Figure SC1. This diagram presents s_a-N and ε_a-N curves for strain-controlled and load-controlled axial-load pertaining to the same material. The smooth shape of the strain-controlled ε_a-N curve is attained in the finite-life region by *suppressing the yielding phenomenon*. Note also that the imposed strains are much larger than any sane mechanical engineer would even consider.

Note that if the load-controlled ε_a-N curves were properly limited to stresses less than the cyclic yield strength, both s_a-N and ε_a-N curves would take on the same shape. Then a load-controlled s_a-N curve might still be preferable to a strain-controlled ε_a-N curve at long fatigue lives because the relevant strain amplitudes might be too small to accurately control.

CHAPTER 16

Strain-controlled Bending, Torsion, and Combined Bending and Torsion Endurance Limits of Round Mild Steel Specimens

I *question the machine design value of mild steel endurance limit data generated using a crank-driven (deflection-controlled) fatigue test machine.* Nevertheless, the fatigue literature contains extensive amounts of such data. The following analysis is based on my compilation of endurance limit data for bending, torsion, and combined bending and torsion, pertaining to over twenty test programs conducted using a *crank-driven, deflection-controlled fatigue test machine with a combined stress attachment.* Note that I am obliged to state my mean stress diagram axes in terms of shear stress to describe the respective endurance limit behaviors, Figure BT1 (top). The slope of the effect of mean stress on endurance limits in this diagram pertain to deflection-controlled mean stress data. (I admit that I am forced to extrapolate this slope to my mean stress diagrams for load-controlled bending and for load-controlled torsion because I do not know of any corresponding endurance limit data in the fatigue literature.)

The actual values for the bending and torsional stresses are not measured *during* crank-driven deflection-controlled combined bending and torsion fatigue tests. Rather the nominal values for the bending and torsional stresses imposed during the first cycle of a given combined bending and torsion fatigue test are established by the values for the imposed deflection and twist at the end of the round cantilever specimen, which in turn are established by the crank eccentricity (throw) and the angular position of the arm of the combined-stress attachment, Figure BT1 *(bottom).* No attempt is made to measure (monitor, observe) cyclic softening behavior during these fatigue tests. Nevertheless, since there is no similar combined stress endurance limit data established by a load-controlled fatigue test machine that I know of, I present certain aspects of my analyses of these combined bending and torsion endurance limits for what it may be worth.

These endurance limit data exhibit an interesting insight to the effect of (i) superimposing torsion on (dominant) bending has no effect in terms of the amplitude for the alternating shear stress pertaining to the endurance limit, and (ii) superimposing bending on (dominant) torsion has a small negative effect in terms of the amplitude for the alternating shear stress pertaining to the endurance limit, and (iii) the respective endurance limit values, each approaching 45 degrees in Figure BT1 from their respective dominant stress sides

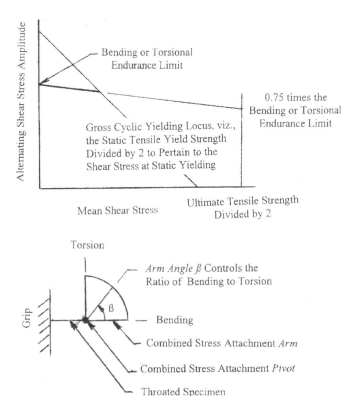

Figure BT1. *Top:* My mean stress diagram for bending endurance limits and for torsional endurance limits of unnotched mild steel specimens. The metric for the respective axes of my diagram is shear stress. Accordingly, normal stresses that pertain to the bending endurance limit, static tensile yield strength, and ultimate strength must be divided by two before plotting on this diagram. *Bottom:* The so-called combined stress attachment is simply an arm that is attached to the cantilever throated specimen at one end and to the connecting rod at the other end. Pure bending occurs at β degrees equal zero and (almost) pure torsion occurs when β is equal to 90 degrees.

do not exhibit a smooth transition (as is typically asserted in mechanics). Rather, a distinct "step" occurs downward (from dominant torsion to dominant bending).

The combined bending and torsional alternating stresses that pertain to the combined stress attachment act in-phase and at the same frequency, viz.,

$$S_{bb,mean} + S_{bb,alternating} \cos(wt)$$

and

$$S_{to,mean} + S_{to,alternating} \cos(wt).$$

A simple method of establishing the amplitude of the alternating shear stress acting on the plane of maximum range of shear stress is to draw *separate* Mohr's circles for alternating stress and for mean stress, but making sure that both circles pertain to the same plane, viz., the plane of maximum range of shear stress. See Figures BT2 and BT3.

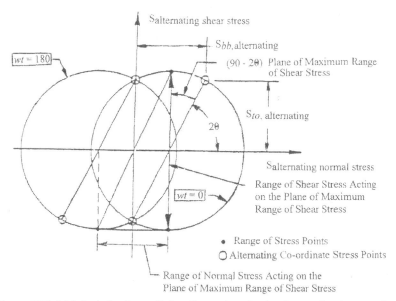

Figure BT2. Mohr's circles for cyclic bending and torsion in-phase and at the same frequency plotted (only when) *wt* equals 0 degrees (the maximum cyclic stress-time history value) and *wt* equals 180 degrees (the minimum cyclic stress-time history value) to establish the maximum value for the *range of shear stress*. The alternating shear stress acting on this plane is one-half of this maximum range of shear stress. The associated mean shear stress acting on the plane of maximum range of shear stress is found by constructing a Mohr's circle for the mean components of the cyclic bending and torsional stresses. See Figure BT3.

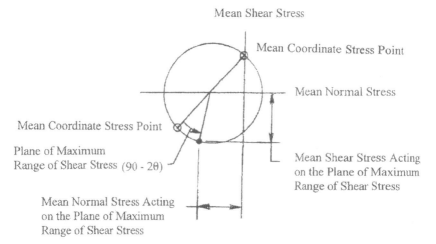

Figure BT3. Mohr's circle for (only) the mean components of cyclic bending and torsion. The mean coordinate stress point in this Mohr's circle is the same coordinate point as the alternating coordinate stress point in Figure BT2. A counter-clockwise rotation of $(90 - 2\theta)$ degrees on this Mohr's circle for mean stress establishes the values for the mean normal stress and the mean shear stress acting on a plane of maximum range of shear stress.

Remark: When the respective Mohr's circles for both mean and alternating stresses are plotted on the same coordinates at time *wt* equals 0 and *wt* equals 180 degrees, the respective maximum ranges of the shear stress and the corresponding ranges of normal stresses are more clearly evident.

CHAPTER 17

Generalized Fatigue Models

ALTHOUGH numerous generalized fatigue models have been proposed in the fatigue literature, none is valid because specific fatigue data exist to refute each model that has ever been proposed. Moreover, such specific fatigue data often existed well before each generalized fatigue model was proposed.

Sines proposed a generalized fatigue model based on the octrahedral shear stress failure criterion for fatigue that included the effect of mean stress, asserting that there is a linear relationship between the (generic) endurance limit and mean stress and, in particular a compressive mean stress increases the (generic) endurance limit whereas a tensile normal mean stress decreases the (generic) endurance limit. In establishing the slope of his linear mean stress line he employed the pulsating tensile and pulsating compressive endurance limit data by Nishihara that is highly questionable because massive plastic deformation occurred during these tests. He then asserted that no endurance limit data existed for superimposed mean normal stress on alternating torsional stress. Accordingly, he could then conduct a test with a compressive mean normal stress superimposed on alternating torsional stress to confirm his model. However, Nishihara published back to back in the bound volume of the Japan Society of Mechanical Engineering his mean stress paper and a paper on endurance limits with *tensile* normal stress superimposed on alternating torsional stress, viz., a test that Sines claimed had never been conducted. Moreover, Nishihara's test *refutes* Sines' model. See Figure GMl. Note that Nishihara' s endurance limit data for the 0.10% plain carbon steel clearly refutes Sines' model which predicts a substantial reduction in the endurance limit with increasing mean tensile stress. The first endurance limit datum value for the 0.72% C plain carbon steel indicates no effect of mean tensile stress, but the two remaining endurance limits datum values

109

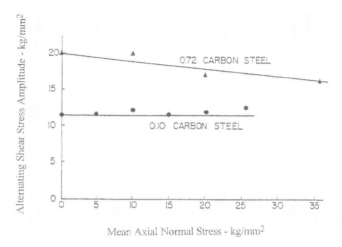

Figure GM1. Nishihara's endurance limit data for tensile mean stress superimposed on alternating torsional stress. (This is just one of so many different fatigue tests that Nishihara conducted. His collection of work is absolutely unexcelled.)

caused Nishihara to plot a mean stress line with a shallow slope, considerably smaller than the slope for the Sines' model. Nishihara later published datum values for a 0.34% plain carbon that is virtually identical to 0.10% plain carbon datum value.

Sines also employed the octrahedral shear stress criterion in his generalized fatigue model. In contrast, I use the shear stress criterion in machine design because it is *safer* than the octrahedral shear stress criterion and also because it is consistent with physical phenomena such a slip, shear lips, and the orientation of Lüder's bands. (However, it is important to understand that even the shear stress criterion is not necessarily safe enough.) The difference between these two criteria is that the shear stress criterion asserts that only the smallest and the largest principal stresses are important, whereas the octahedral shear stress criterion assets that the intermediate principal stress is also important. Findley employed thick-walled cylinders subjected to cyclic internal pressure and cyclic axial normal stresses to conduct fatigue tests with and without a cyclic intermediate principal stress. See Figure GM2. Although his tests on aluminum cylinders pertained only to finite lives, Findley concluded that the intermediate stress did not affect the test outcome. Findley stated that he had previously ran similar fatigue tests on AISI 4340 steel cylinders and obtained data leading to the same conclusion.

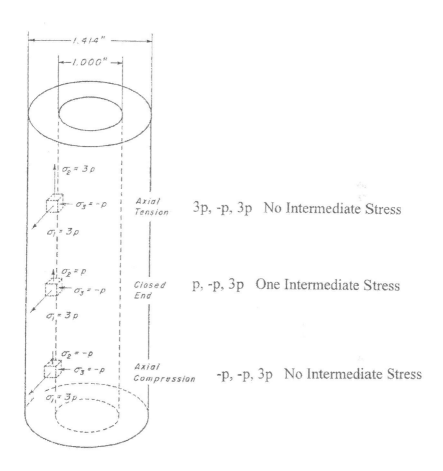

Figure GM2. Findley's thick-walled cylinder specimen that permits running fatigue tests with one intermediate principal stress value and two tests with no intermediate principal value.

CHAPTER 18

The Endurance Limit of Round Mild Steel Specimens under Four-point Rotating Bending with Limited Values of a Superimposed Steady Torsional Moment

STEADY-STATE rotating bending with a superimposed steady torsional moment is the most commonly presumed design state of loading employed the fatigue analysis of mild steel shafts. Accordingly, endurance limits pertaining to this design state of loading was studied early in the history of fatigue testing by Ono (Japan, 1921) and Lea and Budgen (England, 1926). Their results are not only untainted by preconceived notions, but adequately serve to describe the endurance limits of mild steel *provided* the value of steady superimposed torsional moment is appropriately limited.

I summarize below what I believe are the essence of these two publications.

Selected Excerpts from
A. Ono
"Fatigue of Steel under Combined Bending and Torsion"
Memoirs of the College of Engineering
Vol. II, No. 2, 1921
Fukuoka, Japan

Test Machine: Professor Ono built a four-point loading rotating-bending fatigue test machine that also imposed a steady specimen torque induced by an electric absorption dynamometer.

Test Material and Specimens: Ono tested three steels: (1) Series A, a normalized 0.078%C ingot iron (no tension test data given). "The diameter of the test pieces was 6 mm. except two specimens, which were turned down to 5 mm. in order to induce possibly large shearing stress for the given amount of twisting moment." (2) Series B, a normalized 0.304%C ingot iron (no tension test data given). "The diameter of the test pieces was 6 mm. except for a few ones, which were turned down to 4.5 mm. for the same purposes as in the series A." (3) Series D, a quenched and tempered 0.251%C ingot iron (no tension test data given). "The majority of the (test) pieces were 6 mm, in diameter, but 5 (test) pieces were turned to 4.5 mm." Similar to the R. R. Moore four-point rotating-bending fatigue test machine, "The tapered ends of the test piece are tightly fitted in the corresponding conical holes of the specimen holders by means of screws".

However, Ono does not provide a value for the radius blending the cylindrical cross-section of the specimen to its larger tapered ends. Finally, there is no mention of a series C.

Results: Ono does not tabulate his test data but he does plot faired median s_a-N curves with and without steady superimposed torsional shear stresses. He states "According to the results obtained from the series A and B, we learn that the greater part of the (s_a-N) curves for certain values of steady torsional shear stress lies above those for zero steady torsional shear stress; so it appears that the(se) material(s) can withstand greater alternating normal stress in the presence of steady torsional shear stress than in its absence. Further the result of series D shows no considerable influence of steady torsional shear stress on the endurance of the (this) material." (I edited this statement throughout, and below, using the words *steady torsional shear stress*.) "It is to be noted that the capacity of the dynamometer was so small that the steady torsional shear stress could not be made sufficiently large, although the diameter of the test piece was made possibly small in some cases, and therefore the maximum limit of the steady torsional shear stress, which does not produce any bad influence on the endurance, cannot be given."

"As to the broken specimens we may add that fracture occurred in most case along the cross sectional plane near the end of the cylindrical part of the (test) piece, and no remarkable difference was observed between a specimen subject to no torque and that under the said action. Further, if the fracture at the end hastened the breaking, the influence appears insignificant, as the result is not discordant with that obtained from a (test) piece broken in the middle part." (Ono seems slightly apologetic about his cylindrical specimen, as he should be because he is familiar with Wöhler's earlier fatigue tests employing specimens with and without a fillet.)

Finally, in his Summary, Ono stated that "it was proved, though not very exactly, that the endurance(s) of (the) material(s) (tested) under alternating bending (normal stress) is independent of the (superimposed) action(s) of (steady) torsion(al) (shear stress) within the limits of the present experiment." (I added the words and letters in parentheses.)

> **Selected Information Gleaned from**
> F. C. Lea and H.P. Budgen
> "Combined Torsional and Repeated Bending Stresses" Engineering (London), Vol. CXXII, July-December, 1926

Test Machine: The Lea and Budgen test machine differs from Ono's test machine primarily in that the capacity of its steady torsional moment is much larger and secondarily that it (i) employs collet grips and (ii) that its specimen has a 2.5 inch radius at each end of its 0.28 inch diameter cylindrical cross-section.

Test Material: Lea and Budgen also ran test on three steels: (1) A nickel-chrome steel with 0.35%C (no tension test data or heat treatment given). (2) A 0.14%C steel (no tension test data or heat treatment given) (3) A 0.32%C steel (no tension test data or heat treatment given).

Results: Lea and Budgen both tabulate and plot their test data. (1) The superposition of a steady torsional stress equal to 15.2 tons per square inch on the alloy steel specimens barely reduced their endurance limit and the superposition of a steady torsional shear stress equal to 25.1 tons per square inch only slightly reduced their endurance limit. However, the superposition of a steady torsional shear stress equal to 26.2 tons per square inch reduced their endurance limit by approximately one-half. Clearly Lea and Budgen have found the limiting value for the superimposed steady torsional shear stress on the endurance limit of their alloy steel specimens. Given the two plain carbon steels, each demonstrated an increase in the endurance limit when a relatively small steady torsional shear stress was superimposed on the respective sets of specimens. (2) A superimposed steady torsional shear stress equal to 15.1 tons per square inch on the 0.14%C plain carbon steel specimens increased their endurance limit by approximately 5%, whereas a superimposed steady torsional shear stress equal to 19.4 tons per square inch decreased their endurance limit by approximately 15%. (3) A superimposed steady torsional shear stress equal to 15.1 tons per square inch increased the endurance limit of the 0.32%C plain carbon steel specimens by approximately 5% whereas a superimposed steady torsional stress equal to 17.8 tons per square inch decreased their endurance limit by approximately 5%.

Remark: The Lea and Budgen alloy steel data indicating the sharp drop in endurance limit when the value for the superimposed steady (mean) torsional shear stress exceeds a critical values is consistent with the effect of mean stress on axial-load endurance limits.

My Conclusion: The maximum value of the steady superimposed torsional shear stress on the rotating bending normal stress that has a negligible negative effect on the endurance limit should be limited so that the shear stress acting on the plane of maximum shear stress under combined rotating bending and static torsion is kept below the maximum shear stress at yielding. I recommend using the cyclic yield strength rather than the static yield strength as this limiting value for the shear stress at yielding, viz., using 0.9 times (static tensile yield strength)/2 to approximate the cyclic yield strength.

Remark One: The respective data by Ono and by Lea and Budgen ostensibly pertain to specimens that are generically termed unnotched, viz., with a negligible stress concentration. But, because all machine shafts have stress concentrations such as fillets and grooves, we must appropriately adjust unnotched data to account for the relevant generic notch. The problem is that I do not know of any corresponding notched specimen data in the fatigue literature. Thus the attached mean stress diagram is based on my conjecture of what actual notched endurance limit data would confirm.

Remark Two: Provided that the maximum shear stress established by a Mohr's circle pertaining to both alternating bending and mean torsion is less than the static yield strength stated in terms of shear stress, the individual Mohr's circles for alternating bending and mean torsion indicate that the *mean shear stress acting on the plane of maximum range of shear stress is equal to zero.*

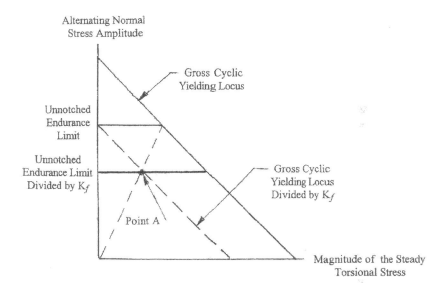

Figure OLB1. Conjectured mean stress diagram for notched rotating bending specimens with superimposed steady torsional stress. Based on my experience with estimating the endurance limit of notched mild steel specimens and the limited notched endurance limit data pertaining to superimposed mean normal stresses, I am confident that my conjectured mean stress locus for a notched specimen with superimposed steady torsion is safe beyond Point A, all the way to the gross cyclic yielding locus. In fact, I suspect that constrained yielding would permit extending this locus even further, but I am reluctant to conjecture just how much further.

CHAPTER 19

Endurance Limit Notch Sensitivity of Round Mild Steel Specimens Under Axial-loading or Bending

FIGURE Q1 depicts typical rotating-bending s_a-N curves for notched and unnotched mild steel specimens. Extensive data of this type were generated between the 1930's and the 1960's using R. R. Moore fatigue test machines. Despite the speed of these machines, the number of laboratory specimens tested was typically only six to eight, with one or two runouts, at most. Note the notch-strengthening effect due to constraint at the root of the circumferential notch is such that the notched and unnotched curves cross at some finite value of fatigue life. Thus it should be exceedingly obvious that the effect of a notch is difficult to explain either in terms of s_a or *fnc*. In fact, the only sensible comparison available pertains to their respective endurance limits. Experience demonstrated that the ratio of the unnotched to the notched endurance limits displayed some consistency, but was typically smaller than the so-called theoretical stress concentration factor, K_t, (where t allegedly connotes theoretical but in fact is almost always experimentally based). Eventually enough data could be compiled to support the endurance-limit notch sensitivity expression

$$K_f = (S_{unnotched\ endurance\ limit})/(S_{notched\ endurance\ limit})$$
$$= 1 + (K_t - 1)\, q$$

in which q is properly termed the ***endurance limit*** *notch sensitivity index* and its value lies between zero and one, *by definition*. Kulm and Hardrath compiled endurance limit K_f data and using an expression given by Neuber provided sufficient information for me to compute the endurance-limit notch sensitivity index curves in Figure Q2. Note that the value for q depends both on the value of radius at the root of the notch and on the value for the ultimate tensile strength of the steel.

The expression for notch sensitivity has a profound analytical consequence. Recall that in stress analysis, we *multiply* the nominal stress (the stress computed using strength of material expressions) by the so-called theoretical stress concentration factor K_t to compute the maximum (local) normal stress at the root of a notch. Then to prevent local yielding, we make sure the maximum local stress is less than the yield strength. In contrast, in fatigue analysis we *divide* the endurance limit by the endurance limit reduction factor, K_f.

Remark One: The notched endurance limit computed (estimated) using this endurance limit notch sensitivity expression is reasonably accurate given the specific data compiled by Kuhn and Hardrath. However, it must be understood that the respective notched and unnotched endurance limits compiled by Kuhn and Hardrath pertain to the same investigator, same test machine, same batch of material, etc. Thus, given only an estimated endurance limit or even a data-based endurance limit, the notched endurance limit computed (estimated) using this notch sensitivity expression is unlikely to be as accurate.

Remark Two: Figure Q2 can also be used to estimate notched axial-load endurance limits given unnotched axial-load endurance limits, even through axial-load data underlying Figure Q2 is much more limited.

Remark Three: Shortly before I dropped out of ASTM Committee E-9 on Fatigue, the definition of K_f was changed to pertain to the ratio of unnotched to notched median fatigue strengths for any *fatigue life* of interest. I opposed this change and attempted to attach enough footnotes so that any newcomer to fatigue who was reasonably intelligent would recognize the *complete irrationality* of such a ratio. (Evidently the change in the definition of K_f was intended to extend the design process to *finite life* analyses, which is utter nonsense. I pity the poor fool who thinks he can design for at least a given median fatigue life (with fifty per cent statistical confidence.).

Remark Four: It is clear in Figure Q1 that considering finite life region, the new so-called finite life K_f value between the respective knees increases above the K_f value that properly pertains to the ratio of endurance limits, and then decreases to the value one at the intersection of the two median s_a-N curves (where q is ostensibly equal to zero), and then takes on a value less than one in the fatigue life region where notch strengthening is evident, resulting subsequently in **negative q** values. Moreover, it is well known that axial-load fatigue tests employing two sets of notched mild steel specimens, one with fillets at its ends and one with a circumferential notch at its center, such that both laboratory specimen geometries have *the same stress concentration factor*, will generate similar endurance limits (and q values) whereas their respective s_a-N curves cross at some finite fatigue life due notch strengthening, resulting in **negative q** values for the circumferentially notched specimens at relatively short finite fatigue lives. Specimens with fillets, of course, do not experience notch strengthening and their finite lives are limited only by the cyclic yielding. Recall Figure U1 in my topic "Ugly s_a-N Curves".

Remark Five: Figure Q3 illustrates that the slope of the downward-sloping, straight-line, finite-life segment of the s-N curve is very strongly dependent on the type of machine employed in testing, whether it is axial-load, rotating bending, or crank-driven plane bending. This dependency clearly profoundly affects the associated value of the new definition of the so-called finite life K_f. Hopefully it is clear that there is no rational finite fatigue life analysis despite what you may find in text books or in so-called research papers or reports. The naiveté of academics who have never ran a fatigue test is profound.

Remark Six: K_f is not simply the ratio of two numbers (values). It is the ratio of the *medians* of two associated statistical distributions. However, the fatigue literature contains endurance limit data for both notched and unnotched mild steel specimens where the respective standard deviations of the presumed endurance limit distributions, if actually computed (estimated), would be greater than one, less than one, or approximately equal to one. Thus the only ratio that makes any sense pertains only to the respective medians, and not to any other percentiles of the respective distributions. This reality clearly negates the notion of a rational reliability fatigue effect, even for endurance limits.

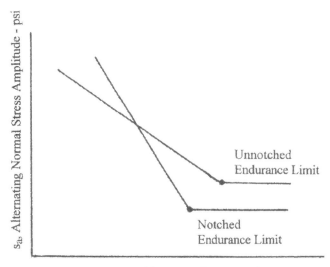

Figure Q1. Typical rotating-bending s_a-N curves for notched and unnotched round mild steel specimens. Note that the knees in the unnotched and notched s_a-N curves pertain to different values of N_{knee}.

Remark One: If while in the library browsing for machine design textbooks, you find one that asserts either (i) both s_a-N knees occur at the same numbers of fatigue cycles, (ii) that both notched and unnotched ultimate tensile strengths are equal, I recommend that you immediately close the book and look for a another machine design textbook. Do not be discouraged if you close most, if not all of the books in the library.

Remark Two: Hopefully it is clear that in this topic I use the notation "N, Fatigue Life - normal stress cycles" and "s_a, Alternating Normal Stress Amplitude - psi" in preference to "N, Fatigue Life - cycles" and "s_a, Alternating Stress Amplitude - psi" to emphasize *normal stress* as opposed to *shear stress*. However, in design analysis the appropriate stress metric is the metric used in the failure criteria ultimately employed in designing against fatigue failure. I recommend using the *maximum shear stress criterion* instead of the octahedral shear stress criterion. See my topic "Generalized Fatigue Models".

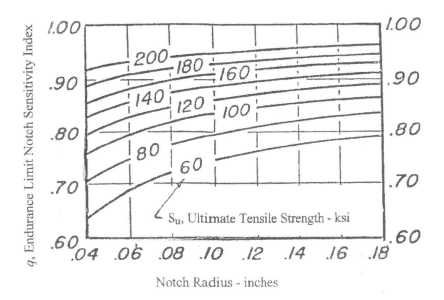

Figure Q2. Curves for the endurance limit notch sensitivity index, q, pertaining to mild steels only, computed using information found in Kuhn and Hardrath (NASA TN2805, 1951).

Remark: Notched and unnotched steel fatigue specimens hardened to have ultimate tensile strengths greater than about 220 ksi usually do not exhibit a distinct knee in their respective s_a-N curves, but nevertheless these s_a-N curves become horizontal for practical purposes at very long fatigue lives. Notch sensitivity index values, q, typically are often so close to 1.0 that K_t is the same K_f for practical purposes. Notch strengthening is no longer exhibited at these ultimate tensile strengths (hardness values) and in some cases the respective unnotched and unnotched rotating-bending s_a-N curves are almost siblings in appearance.

Caveat: Axial-load fatigue tests conducted on unnotched and notched steel specimens with ultimate tensile strengths above 220 ksi require much more attention given to preventing superimposed bending stresses due to poor alignment.

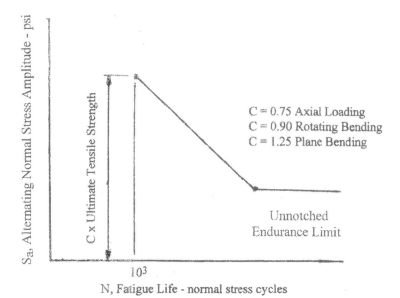

Figure Q3. A sketch of the two straight-line segment s_a-N curve for mild steel fatigue specimens indicating that the slope of the downward-sloping, straight-line finite-life segment of the s_a-N curve is very strongly dependent on the type of machine employed in testing, whether it is axial-load, rotating bending, or crank-driven deflection-controlled plane bending. The C values given in this sketch are based on extensive data, particularly for rotating-bending tests and for crank-driven deflection-controlled plane-bending tests.

Remark: Recall my topic "Ugly s_a-N curves". Any unnotched s_a-N curve that involves massive micro-yielding is ugly in a design context. The only design purpose of an unnotched fatigue test is to provide a basis for "correcting" (modifying) the unnotched s_a-N curve to account for the notch that is relevant to the design application of specific interest. Any such "correction" (modification) methodology must pertain to material behavior that can reasonably be regarded at least as quasi-elastic so that the associated strength of materials nominal stress expressions can be regarded as at least quasi-credible.

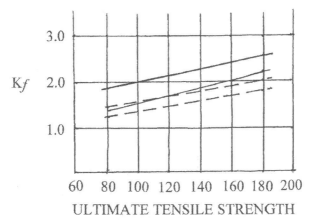

Figure A1. Conjectured K_f values for keyways. The solid lines pertain to alternating bending, whereas the dashed lines pertain to alternating torsion. The higher paired lines pertain to profile keyways, whereas the lower paired line pertain to sled-runner keyways. (Use the sled-runner curves for Woodruff keyways.)

19.1. APPENDIX A

It should be clear that the endurance limit reduction factor K_f cannot be computed for sled-runner and profile keyways using the standard text expression for K_f. (In fact, this inability pertains to all complex geometrical shapes.) Nevertheless, because these keyways are universal, it is common practice to employ a *conjectured* K_f value in design. Accordingly, I dug out Figure A1 from my 1960's files. Unfortunately, I don't remember enough about my development of Figure A1 conjectured values to assert today that this diagram is reasonably accurate. I hope so, but I just don't know.

What I do know is that the *product* of K_f and the naive notched endurance limit factor of safety should, if possible, have the same value for all notches. Accordingly, I recommend that (i) always use a sled-runner keyway whenever practical, and (ii) if possible, only use a profile keyway inside a press fit.

CHAPTER 20

The Effect of Mean Stress on Load-controlled Axial-load Endurance Limits of Round Mild Steel Specimens

A number of so-called mean stress lines have been proposed in the fatigue literature to describe the effect of mean stress on the axial-load endurance limit of mild steel. None are accurate. The best known study of the effect of mean stress on the axial-load endurance limit of mild steel specimens is due to J. O. Smith in 1940. He employed a plot of the modified Goodman diagram on Haigh-Soderberg coordinates (Figure MS1) to display his compilation of axial-load mean stress endurance limit datum values. He concluded that the modified Goodman diagram safely represents axial-load endurance limit data for mild steels. However, (i) he was evidently unaware of a common sense 1933 German industrial standard that modified the modified Goodman diagram to pertain only to maximum cyclic stress values less than the static yield strength of the given mild steel. Moreover, (ii) he had *previously* conducted load-controlled torsional mean stress tests and observed massive *permanent* rotational deformation when the maximum cyclic stress exceeded the static torsional yield strength. Accordingly, he subsequently limited his mean stress values to only maximum cyclic stresses below the static torsional yield strength. But this constraint limited his valid mean stress test values so drastically that he concluded that his remaining (valid) data indicated *no effect* of mean stress on the torsional endurance limit of (unnotched) mild steel specimens. Inexplicability, (iii) he did not then apply the same maximum cyclic stress constraint to properly edit his compilation of axial-load endurance limit datum values. Unfortunately, Smith published his unedited compilation. His modified Goodman diagram is not only overly safe, it has the *wrong* slope.

The axial-load endurance limit datum values for (unnotched) mild steel specimens that I used to construct my mean stress diagram were

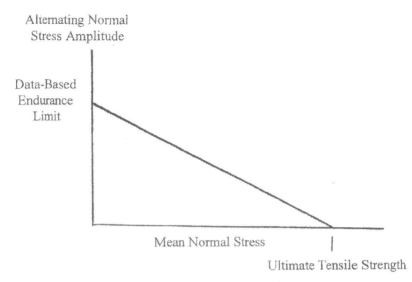

Figure MS1. Smith plotted his compilation of endurance limit data for mild steels using Haigh-Soderberg coordinates for his modified Goodman diagram. However, he should have known better than to include endurance limit datum values that pertained to maximum cyclic normal stresses above the static yield strength. He had previously run mean stress tests in torsion in which he had limited the maximum torsional stress to less than the torsional yield strength to prevent markedly excessive permanent deformation. If he had compiled axial-load endurance limit mean stress data of interest in machine design, he would have generated a plot similar to that in Figure MS2.

limited to maximum cyclic stresses that did not exceed my estimate of the associated *cyclic* yield strength, viz., 0.9 times the static tensile yield strength. See Figure MS2. My mean stress diagram is presented in Figure MS3. In turn, my modification of this mean stress diagram to account for the effect of notches is presented in Figure MS4 and Figure MS6, with supporting information provided in Figure MS5 and Figure MS7.

Remark One: When the respective axial-load mean stress endurance limits data plots for mild steel specimens are examined individually, the most striking observation is that when the maximum cyclic stress reaches the static yield strength. The endurance limit data often suddenly drops and follows along the static yield strength locus.

Remark Two: R. M. Brown was a university professor in Scotland who ran mean stress axial load endurance limit tests on mild steel specimens. He found that a power fluctuation would cause high mean stress specimens to neck down and fail after running for a considerable period of time without incident before the power fluctuation. He subsequently ran his tests only in the summertime when classes were not in session and power fluctuations did not occur.

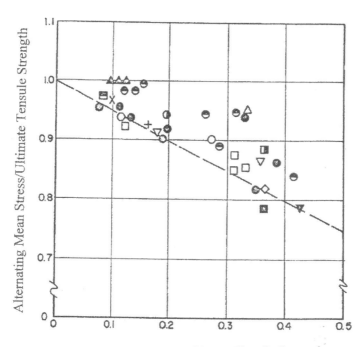

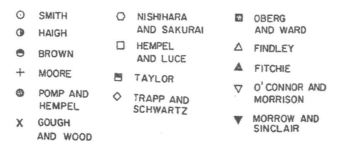

Figure MS2. The (dashed) mean stress line pertains to axial-load endurance limits for unnotched mild steel specimens based on an *edited* compilation of published mean stress data for which the maximum cyclic normal stress was less than the associated cyclic yield strength (estimated to be approximately 0.9 times the static tensile yield strength). The dashed line is adequate, in my opinion, to represent a credible machine design endurance limit criterion.

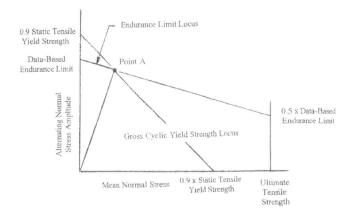

Figure MS3. My mean stress diagram for the axial-load endurance limits of unnotched mild steel specimens. This diagram is subsequently modified to pertain to notched specimens by treating the endurance-limit reduction factor K_f as being equivalent to a stress concentration factor in the region to the left of the radial line joining the origin and Point A. Accordingly, the resulting notched endurance limit locus in this region is parallel to the unnotched endurance limit locus, Figure MS4.

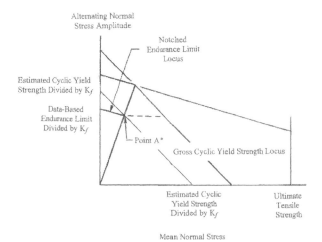

Figure MS4. Modification of my mean stress diagram for endurance limits of unnotched mild steel specimens to account for effect of notches by treating K_f as if it were a stress concentration factor. Accordingly, Point A* pertains to local cyclic yielding. Then, for larger imposed normal stresses, the Gunn model asserts that (i) the local maximum normal stress at the root of the notch cannot exceed the (local) cyclic yield strength and therefore (ii) regardless of the imposed value for mean normal stress, the *range of normal stress at the root of the notch is a constant.* Hence the dashed horizontal extension of the notched endurance limit locus. The problem with Gunn's model is that it predicts higher endurance limits that the limited existing endurance limit data for mild steel specimens. I think this problem is likely due to his equating the local cyclic yield strength to the gross cyclic yield strength, viz., he ignores *constrained yielding* at the root of the notch. See Figure MS5.

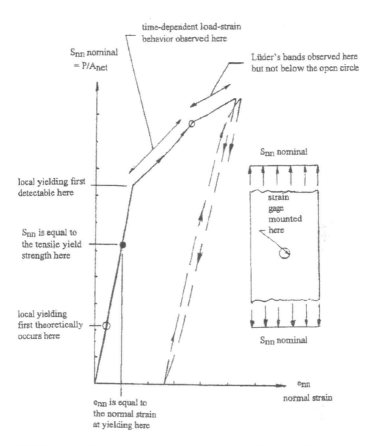

Figure MS5. Outcome of a tension test that I conducted employing a notched specimen that illustrates the notion of constrained yielding at the root of a notch. The specimen material was hot-rolled 1008 plain carbon mild steel, one-eighth of an inch thick, five and one-half inches wide with a one inch diameter hole at its mid-width. A thirty-second of an inch long foil strain was mounted longitudinally on the *inside surface* of the hole. The imposed normal load P in pounds was the independent variable and the associated normal stain e_{nn} was the dependent variable (e_{nn} is established by the change of resistance of the foil strain gage.) Previous conventional (unnotched) tension tests established the value for the normal strain e_{nn} at yielding. The loading portion of the notched tension test was linear (as well as could be determined by examining the individual P versus e_{nn} datum values) to the point where local yielding clearly occurred with a sudden obvious deviation from the previous linear behavior. After Lüder's bands were observed the specimen was unloaded and reloaded numerous times, generating hysteresis loops that continually ratcheted slightly to the right. In the time dependent of the load-strain plot, a sustained load exhibited an increase in strain due to *local stress redistribution.*

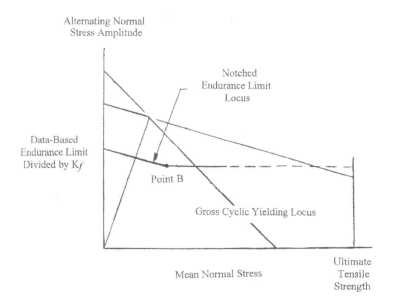

Figure MS6. Simplified notched endurance limit locus. I recommend extending the notched fatigue locus to Point B before further extending it horizontally, where Point B is approximately half-way between the nominal elastic region in Figure MS4 and the gross cyclic yield strength locus. This simplified methodology establishes a locus that lies below all mean stress notched endurance limit data that I have seen. But for critical design applications, extending the downward sloping portion of this locus to the gross cyclic yield strength locus before converting to its horizontal portion is surely a much more conservative methodology. Finally the limit on the horizontal extension of this locus depends on notch strengthening effect if any for the given specimen. See Figure MS7. The horizontal portion can sometimes be extended to the point where the mean stress value is equal to or greater than the material ultimate tensile strength. (Nevertheless, I recommend limiting the maximum value of the imposed mean stress as a matter of principle to minimize possible permanent deformation.)

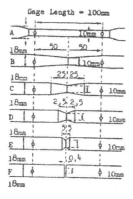
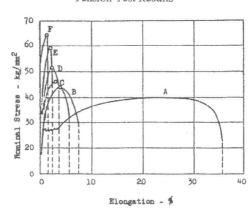

Figure MS7. Ludwig and Scheu tension test data for an unnotched and for five circumferentially notched specimens mild steel specimens. Note that more severe notches exhibit greater notch strengthening due to greater constraint at the root of their notches. If the notch is sufficiently severe a ductile (growth and coalescence of voids) crack initiates at the root of the notch and propagates inward until the remaining uncracked cross-sectional area fractures abruptly with its portion of failure surface exhibiting (brittle) cleavage. The boundary between ductile and brittle failures appears very distinct to the eye, but a SEM micrograph shows that there is actually a region of mixed growth and coalescence and cleavage. On the other hand, if the notch severity is only slightly greater than that of a neck in an unnotched tension test, the crack initiates internally and very rapidly grows radially outward until fracture occurs abruptly when the circumferential shear lips are formed. This so-called cup and cone failure is a mixed mode of ductile (growth and coalescence of voids) failure with the relatively flat cup portion failing under plane strain (constraint) conditions whereas the cone portion is formed shear lips under plane stress conditions and exhibit and elongation of the voids in the direction of sliding. In both cases, however, fatigue cracks originate at the root of the notch and propagate radially inward until abrupt fracture occurs.

20.1. APPENDIX A

The endurance limits of mild steel specimens depend so critically on the alignment of the specimen grips for each given fatigue test machine that it is impossible to provide a single *generic design curve* for the endurance limit of mild steels under axial-loading. Haigh in about 1930 recommended estimating the axial-load endurance limit as 0.6 times the rotating-bending endurance limit. This ratio was based on endurance limit data generated using his Haigh Pulsator axial-load fatigue test machine that he designed, built, and sold commercially. By 1940 Fairies' machine design book recommended using a ratio equal to 0.7, which subsequently changed to 0.85 by 1950. Some contemporary machine design textbooks recommend even higher ratios. The fact that the alignment associated with commercial axial-load fatigue test machines has improved markedly over the years does not mean that your specific design will ever attain similar alignment. I provide two generic design curves in Figure MSA: one for very precise alignment and one for typical (good) alignment, where typical (good) refers to most contemporary commercial fatigue machines, not to typical alignment in service and not necessarily to the alignment associated with your specific design.

I do not believe you ever get very precise alignment using threaded grips, even though rolled or ground male threads are surely better than chased or die-cut male threads. I know that ground button-head specimens with a ground split-collets provide much better alignment in a tension test. Perhaps some analogous attachment may improve the alignment in the design of specific interest.

Regardless of the type of gripping, its tightness and its rigidity, the alignment is always better when the imposed tensile mean stress is increased. This is particularly true for sheet steel. I have never run a fatigue test on a sheet steel specimen that did not have its mean normal stress (at least slightly) greater than the imposed alternating normal stress. In fact, in my topic "Endurance Limits of Mild Sheet and Plate Steels", I say "Axial-load tests on sheet and plate should always involve a tensile mean normal stress slightly greater that the imposed alternating normal stress to *avoid buckling*, but with the maximum tensile normal less than the cyclic yield strength *to avoid excessive plastic deformation.*" This means that it may not be possible to generate credible endurance limit data for some annealed and normalized low-carbon *unnotched* steel specimens. See Figure SP2 in that topic.

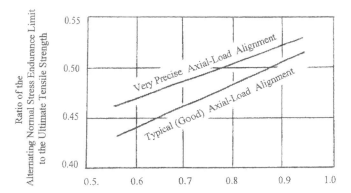

Figure MSA. A plot of my *generic design curves* for axial-load endurance limits pertaining to round mild steel specimens as established by contemporary commercial fatigue test machines (good alignment) and by fatigue tests in which experienced *extra care* is given to improving the alignment of contemporary commercial fatigue test machines (very precise alignment).

20.2. APPENDIX B

Figure MSA provides a preliminary estimate of the axial-load endurance limit for plotting on the $S_{alternating}$ ordinate in Figure MSB. The maximum value for S_{mean} that still pertains to the nominal elastic region for unnotched specimens is established as follows, presuming $S_{cyclic\ yield\ strength}$ equals $0.9 S_{yield}$:

In turn, using similar triangles the corresponding maximum value for S_{mean} for notched specimens is easily calculated, given the estimated value for K_f

$$S_{alternating} + S_{mean} = 0.9 S_{yield}$$

$$S_{alternating} = S_{Endurance\ Limit} - \left[\frac{S_{Endurance\ Limit}}{2}\right]\left[\frac{S_{mean}}{S_{Ultimate\ Tensile\ Strength}}\right]$$

Thus

$$S_{mean} = \frac{0.9 S_{yield} - S_{Endurance\ Limit}}{\left[1 - \frac{S_{Endurance\ Limit}}{2 Ultimate\ Tensile\ Strength}\right]}$$

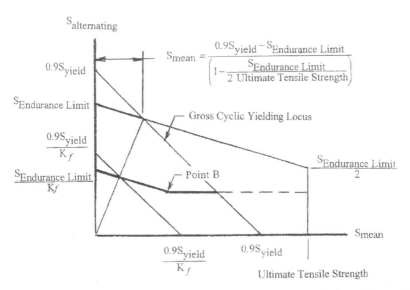

Figure MSB. My load-controlled axial-load mean stress diagram for notched round mild steel specimens.

CHAPTER 21

Axial-load Endurance Limits for Mild Steel Sheet and Plate Specimens

MILD steel sheet and plate specimens are typically fatigue tested in tension using a load-controlled fatigue test machine or in bending using a deflection-controlled (strain-controlled) crank-driven fatigue test machine. Axial-load unnotched specimens typically have a streamlined geometry with radii as large as practical, usually terminating at the sides of the specimen at the respective starts of its parallel grip regions. See Figure SP1(a). This specimen clearly requires shims to restrain it from buckling unless the mean axial normal stress exceeds the alternating axial normal stress. Plane-bending unnotched specimens typically have the so-called *constant stress* geometry displayed in Figure SP1(b).

I could recommend that unnotched mild steel sheet and plate fatigue specimens be tested only in axial-loading conditions where its imposed mean normal stress is always somewhat greater that the imposed alternating normal stress, but with its maximum cyclic stress smaller than the gross cyclic yield strength. However, these two loading conditions are so restrictive that it typically is not possible to generate credible endurance limit data for annealed, hot-rolled or normalized low-carbon steel sheet or plate specimens. See Figure SP2. Thus, for machine design components fabricated from low-carbon sheet and plate, I cannot advocate attempting to find published fatigue data that emulates the service conditions for the given design application. The differences between the geometry and loading of sheet and plate fatigue specimens and the respective geometries and loadings of even the simplest machine design sheet and plate components are too extreme to allow credible extrapolation.

The methodology that I recommend to estimate sheet and plate endurance limits for notched mild steel specimens is to construct a con-

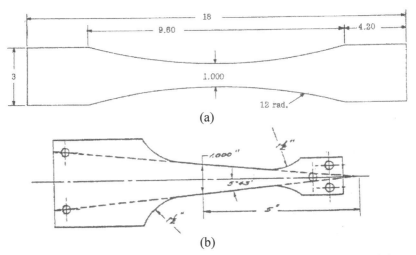

Figure SP1. (a) The axial-load fatigue test specimen geometry pertaining to the fatigue test data in Figure UT1 in the topic "The Ugly Truth About Fatigue Tests". (b) Sheet and plate plane-bending specimen with the so called *constant stress* geometry.

jectured mean stress diagram for unnotched mild steel sheet and plate specimens using my mean stress axial-load diagram in the topic "The Effect of Mean Stress on the Load-Controlled Axial-Load Endurance Limit of Mild Steels" in conjunction with my *generic design curve* (Figure MSA) for good alignment depreciated by fifteen percent for machined edges and by thirty percent for sheared edges. Then, (i) construct the associated conjectured notched endurance limit locus, given your estimate of the relevant stress concentration factor in the given design application of interest and (ii) using the endurance limit notch sensitivity index value in Figure Q2 in the topic "Endurance Limit Notch Sensitivity of Round Steel Specimens". I regard this conjectured locus to be reasonably credible for any combination of imposed alternating and mean normal stresses at any location in any design application of interest, where experience (or common sense) indicates that either gross or local buckling is not an obvious problem, e.g., for tubular members, stamped brackets with beads, welded frames and housing, especially those with gussets and reinforcements. Accordingly, the primary differences between my estimated axial-load endurance limits for notched round specimens and my conjectured axial-load endurance limits for notched sheet and plate specimens is (i) the absence of a notch strengthening effect for sheet and plate specimens and (ii) my depreciation factors for sheet and plate specimen edge conditions.

In designing brackets and small housings with sheet and plate, I recommend first performing minimal sizing calculations in your synthesis stage. Then when the design appears feasible and practical, re-visit your minimal calculations and make any obvious changes required. In turn, alleviate your remaining concerns with local support and stiffening rather than increasing the overall size or the gage of the sheet or plate.

Remark: If a sheared edge of sheet or plate is subsequently bent to form a flange of some type, try to make sure that the "cut and smear" side of the sheared edge experiences tension and the "tear" side experiences compression during bending.

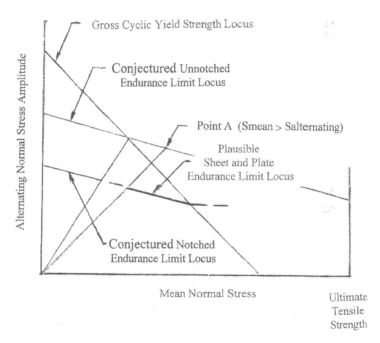

Figure SP2. My conjectured mean stress diagram for the endurance limits pertains to axial loading of unnotched low-carbon sheet and plate specimens. I am obliged, because of the lack of credible data, to employ the generic design curves in Figure MSA in my topic "The Effect of Mean Stress on the Load-Controlled Axial-Load Endurance Limits of Round Mild Steel Specimens" as surrogate data. Accordingly, both the unnotched and notched endurance limit loci are constructed as described in that topic. However the respective loci are conjectured. Nevertheless, the solid line portion of the notched endurance locus that lies to the right of the line joining the origin and Point A is at least plausible. Moreover, for plate this locus may be extended horizontally somewhat, depending on the amount of constrained yielding, if any, that occurs for the given notch.

CHAPTER 22

Torque-controlled Torsional Endurance Limits of Round Mild Steel Specimens

J.O. SMITH ran mean stress endurance limit tests in torsion that might be regarded as torque-controlled in a cyclic-deformation context. His mean-stress endurance-limit specimens actually exhibited very large *permanent* twist angles. Smith conceded that the endurance limit is unimportant when it is accompanied by unacceptable values of permanent deformation that could only be avoided by limiting the maximum value of the cyclic shear stress imposed during fatigue testing.

Smith concluded that mean stress has no effect on the endurance limit for unnotched specimens, but that there is an effect of mean stress on the endurance limit for notched specimens. (This dumbfounding conclusion makes no sense.)

I assert that the appropriate upper limit for the cyclic shear stress pertaining to torque-controlled endurance limits is the actual cyclic torsional yield strength that pertains to *hollow* specimens. Typically for mild steels the ratio of hollow specimen torsional yield strength to the conventional 0.2% offset yield strength for solid specimens is about 0.80 to 0.85. In turn, the cyclic torsional yield strength for solid specimens is about 0.8 times the corresponding static torsional yield strength. Thus I assert that the upper limit of the maximum cyclic torsional stress for solid unnotched specimens is about 0.65 times the conventional static torsional yield strength. This limit is so severe that I believe very few, if any, torque-controlled mean-stress unnotched torsional endurance limit values are actually credible in machine design applications. Nevertheless, I opt to construct the slope of the mean stress unnotched torsional endurance limit locus as presented in Figure TCT1, where the unnotched torsional endurance limit is conservatively estimated as 0.5 times the R.R. Moore rotating bending unnotched endurance limit in Figure RRM4 in topic *R. R. Moore Four-Point Rotating Bending Fa-*

tigue Test Machine. The resulting unnotched specimen torsional endurance limit locus is fictitious for practical purposes. Its only use is to construct the conjectured notched specimen endurance-limit locus in Figure TCT1.

The remaining issue is to establish a reasonable value for K_f. Round mild steel specimens are less sensitive to circumferential notches under torsional loading than under axial-loading or bending. Accordingly, I think that a reasonable estimate of the value for the endurance limit notch sensitivity index q in torsion is about 0.85 times the value for the q value for axial-loading or bending.

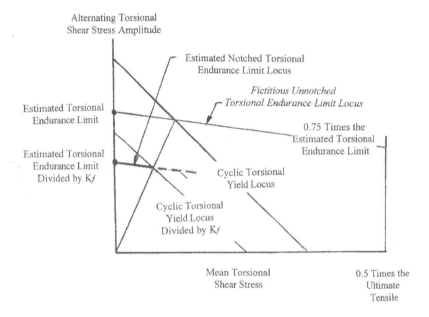

Figure TCT1. My Conjectured Torque-Controlled Notched Torsional Endurance Locus. The dashed portion of this locus pertains to constrained yielding at the root of the notch. It surely is safer to limit this locus to just the nominally elastic region (with no dashed extension for constrained yielding).

CHAPTER 23

Fatigue Factor of Safety

PERSPECTIVE: The following discussion of a naive factor of safety against fatigue involves endurance limit mean stress diagrams, Mohr's circles for alternating and mean stress, critical cross sections and critical stress elements, load lines, etc. This topic may not be of interest if you already have access to service proven design procedures. On the other hand, if you are trying to establish a new design procedure for some machine element, then I recommend that you read this topic to see if it gives you any ideas.

Background: The generic factor of safety that I refer to as a naive factor of safety is simply the ratio of the design stress to the associated material strength. It is based on the intuitive notion that to prevent failure, the stress pertaining to an unspecified mode of failure must be less than the strength pertaining to the associated unspecified mode of failure. (Recall that strength is merely the particular value of stress in some well-defined test protocol which establishes an event that is commonly referred to as failure.) In a broader sense, stress is the stimulus for failure and strength is the resistance to failure.

The design stress is commonly denoted DSS for the *design state of stress*. The associated strength is considered to be the failure stress and is commonly denoted FSS, for the *failure state of stress*. Accordingly, the naive factor of safety is merely the (dimensionless) ratio FSS/DSS. I usually refer to a naive factor of safety *against a specific mode of failure,* especially against yielding or against fatigue. Moreover, the respective stimulus and resistance metrics must be identical and *stated in terms of the presumed failure criterion.* Since I always use the shear stress failure criterion, my proper metric is always *shear stress*. (Recall that for tension only or for bending only, the shear stress is simple one-half of the normal stress.)

Remark: No failure criterion is correct. But the shear stress criterion has more credibility and is safer than the octrahedral shear stress criterion. Unfortunately its domain of credibility is limited due to the underlying presumption of pertaining to an isotropic material. Moreover, in certain situations, even the shear stress criterion is not safe enough.

What I call a Method One (textbook) *design factor of safety* attempts to be more credible than a naive factor of safety by introducing a third state of stress, the so-called *stand-by state of stress,* denoted SSS. It is the state of stress when the machine is *not* being operated. A design factor of safety also introduces the concept of a *load line* which depicts how on a *failure locus plot* the actual service loading is presumed to increase from the DSS to the FSS. In fatigue analysis, the DSS is also typically presumed to pertain to a *steady-state* loading condition associated with the anticipated normal operation of the given machine. In contrast, a *sizing factor of safety* pertains to how incremental changes in the size of a potentially critical cross-section of the component moves the plotted DSS point along a path towards its intersection with the failure criterion locus. This intersection establishes the location of the plotted FSS point on the appropriate failure locus. A sizing factor of safety is generally employed in preliminary design to assure that the respective potentially critical cross-sections of each component of specific interest will have a (more or less) uniform factor of safety, subject to certain practical geometric constraints. The load line for a sizing factor of safety is determined the same way for each given cross-section. In contrast, the alternative plausible load lines for a design factor of safety are established by knowledge, experience, comparisons, analogies, worst-case presumptions, or even by hypotheticals. Ideally, service feed-back or experience indicates how different changes in certain loading conditions have previously caused failure. If so, then the respective alternative design factors of safety against these modes of failure have to be related somehow to the probability of each type of failure and to the corresponding intensity of the consequences associated with that type of failure. Typically this complex process is drastically simplified by merely designing against the failure scenario employing the most extreme, but reasonably anticipated, loading conditions.

Remark: Without experience or knowledge regarding the most likely location of the critical cross-section and the anticipated mode of failure; design analysis and its associated ambiguous naive design factor of safety is essentially based on a foundation of guesswork. Actual failures of analogous components in analogous applications can provide useful design information, but actual tests on prototype components provide more credible information. Fortunately, very few designs start from scratch.

Generic Fatigue Example: Figure FFS1 depicts the relevant portion of the notched mean stress endurance limit locus in Figure MSB in Appendix B of the topic "The Effect of Mean Stress on the Axial-Load Endurance Limit of Round Mild Steel Specimens". It is the failure locus of interest in designing against axial-load fatigue failure, *expressed in terms of shear stress.* The fatigue version of what I call the *Method One design factor of safety against fatigue failure* involves three distinct points plotted on this mean stress notched endurance limit diagram: (i) the SSS point, the standby state of stress point, which includes all known residual stresses and steady stresses (associated with dead

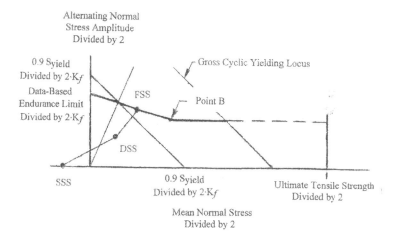

Figure FFS1. The axial-load notched endurance limit locus, *stated in terms of shear stress.* The safest design pertains to load lines completely within the nominal elastic region. Nevertheless, any load line that intersects the notched endurance limit locus inside of the estimated gross cyclic yield strength locus, 0.9 times S_{yield} divided by two, is safe based on limited published data. Even more limited data indicate that the dashed line is safe for notches with relatively large notch strengthening effects. (In fact, I have seen data for large mild steel bolts with an endurance limit that was only very slightly decreased as the mean stress increased from less than one-half of the ultimate strength to more than three halves of the ultimate tensile strength.)

loads and assembly stresses) that are imposed on the *critical stress element* located on the critical cross-section of specific interest prior to the imposition of the design state of stress loads; (ii) the DSS point, the design state of stress point, which includes the SSS stresses plus the externally imposed *allegedly steady-state* DSS stresses acting on the *critical stress element,* and (iii) the FSS point, the fatigue failure state of stress which depends how the individual DSS stresses acting on the critical stress element are alleged to increase from the plotted DSS point to intersect the notched mean stress endurance limit locus.

For the simplest Method One design factor of safety against fatigue failure: (i) the coordinates of the SSS point are (0,0) on the Haigh-Soderberg mean stress notched endurance limit diagram in Figure FFS1; (ii) the coordinates of the DSS point and (iii) the coordinates of the FSS point, which are established by extending the straight line that passes through both the SSS point and the DSS point to intersect the notched fatigue endurance limit locus. This simplest Method One design factor of safety against fatigue failure is numerically identical to the classic naive factor of safety.

The general form of the Method One design factor of safety against fatigue failure has straight line segments from the SSS point to the DSS point and from the DSS point to the FSS point; where the FSS point is most credible when service-based experience provides actual loading conditions information that depicts how the load line from the DSS to the FSS could occur or actually has occurred for analogous applications. Unfortunately this information is usually not available and accordingly all plausible alternative FSS point locations should be examined to establish to the respective magnitudes of their Method One design factors of safety against fatigue failure. The generic expression for this Method One design factor of safety against fatigue failure is the ratio of the length of the straight line segment for the SSS point to the DSS point divided *into the sum* of the lengths of the straight line segments from the SSS point to the DSS point and from the DSS point to the FSS point. (i) Note that only the SSS and DSS points require a *critical stress* element located on a *critical cross-section* to compute the coordinate stresses used to plot their respective points on Figure FFS1. (ii) Note also these respective critical stress elements are not even constrained to have to be at the *same location* on each potentially critical cross section. Thus I do not regard this textbook Method One design factor of safety against fatigue failure as being technically competent. At best, this textbook analysis is only technically expedient.

Fatigue Factor of Safety **147**

My design factor of safety against fatigue failure, which I call Method Two, is the rational technical, alternative to the Method One design factor of safety against fatigue failure. I assert that the SSS point, the DSS point, and the FSS point *should all pertain to the same critical stress element, at the same location on the same potentially critical cross-section and thus pertains to the same plane and direction within this critical stress element!* This plane of maximum range of shear stress is established by the fatigue Mohr's circles pertaining to the FSS stress element!

Remark: The three respective states of stress, SSS, DSS, and FSS for the simplest form of the Method One design factor of safety against fatigue failure stated above also satisfy the requirements of a Method Two design fatigue factor of safety against fatigue failure because all three states of stress pertain to the same critical stress element on the same (potentially) critical cross-section and to the same plane within each critical stress element.

The general form of my Method Two design factor of safely against fatigue failure is somewhat more circumspect than its Method One counterpart. It starts with the Method One FSS point and its critical stress element, FFS2(a). The plane of maximum range of shear stress and established using Mohr's circles as in Figure FFS2(b). The corresponding mean stress on this plane is established in Figure FFS2(c). In turn, the direction of the alternating shear stress on this plane is expressed in terms of the direction cosines relative to the physically relevant coordinate system for the given component. In turn, given the coordinate alternating and mean stresses for the respective DSS and SSS *pertaining to the FSS critical cross-section and critical stress element,* and also given *the FSS direction cosines relative to the relevant refer-*

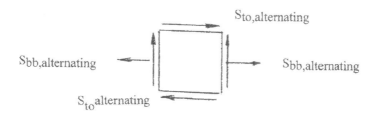

Figure FFS2(a). The typical DSS alternating stress element with (i) combined bending and torsion (shown,) or (ii) only bending or (iii) only axial-loading, or (iv) only torsion.

> I previewed Mohr's circles for combined bending and torsion cyclic stresses in-phase and at the same frequency in my topic "Strain-controlled Bending, Torsion, and Combined Bending and Torsion Endurance Limits of Round mild Steel Specimens". I now repeat Figures BT2 and BT3 in the context of factor of safety analyses.

ence frame, the corresponding alternating and mean shear stress on the FSS plane in the FSS direction can be established, respectively, for the DSS and SSS. Then the expression for the Method Two factor of safety is the same as for Method One, viz., the length of the SSS to DSS segment divided into the sum of the load line segments.

Regardless of the value adopted for the Method Two (or Method One) factor of safety against fatigue failure, it is absolutely vital that adequate prototype testing take place before committing to manufacturing that particular machine component. Overall, I recommend a value for the Method Two sizing factor of safety against fatigue failure of either three and four for "infinite life". I recommend the value of three in applications

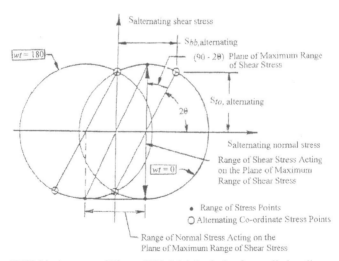

Figure FFS2(b). A repeat of Figure BT2. Mohr's circles for cyclic bending and torsion in-phase and at the same frequency plotted (only when) *wt* equals 0 degrees (the maximum cyclic stress-time history value) and *wt* equals 180 degrees (the minimum cyclic stress-time history value) to establish the maximum value for the *range of shear stress.* The alternating shear stress acting on this plane is one-half of this maximum range of shear stress. The associated mean shear stress acting on the plane of maximum range of shear stress is found by constructing a Mohr's circle for the mean components of the cyclic bending and torsional stresses. See Figure BT3.

where fatigue failure would not cause a serious problem and replacement is typically without further problems. I recommend the value of four in applications where fatigue would be a more serious problem, either in terms of the cost to repair, or the cost of down time, and particularly if failure might cause extensive damage to other components.

Before ending this discussion of factor of safety, I want to emphasize that the primary value of a factor of safety is in its use as a metric to *extrapolate* service proven performance *to* the design of your component Thus it is important to generate and maintain comprehensive records of performance for all analogous designs and to continually test the current factor of safety algorithm regarding its ability to distinguish between successful designs and designs that have exhibited problems.

Remark One: Automotive industry gossip and junk yards provide an automotive designer a lot more design information than any textbook. Side by side comparison of components and products always provides very valuable design information regardless of the specific industry.

Design is much simpler when it pertains to a machine or a machine component that is analogous to an existing machine or machine component *with a known service history.* Then the concept of a *bogey* can be coupled with appropriate testing to extrapolate service proven experience to the future. This methodology is termed *quality assurance* in the automotive industry. The initial bogey tests were developed by J.O. Almen who ran laboratory fatigue tests on certain components in an attempt to generate fatigue failures that occurred at the same location and had the same appearance as test track and junkyard failures. Then, if a particular type of component has passed its specific bogey test each year for several model years, during which the production components have demonstrated satisfactory service performance, it is reasonable to presume that its service performance will be satisfactory.

Remark Two: A factor of safety is not necessary if a service proven algorithm for a design with years of satisfactory performance is available. There is an old axiom: If it's not broken, don't fix it.

Remark Three: If an existing design procedure always works well, do not fall for the argument that it is probably over-designed. If there is an issue with its weight or size, than redesign by proportions might be appropriate. Remember the old axiom.

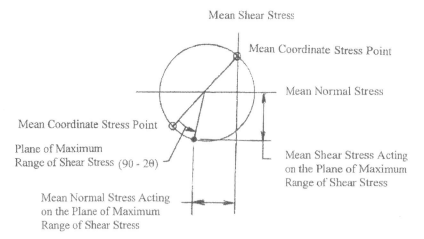

Figure FFS2(c). A repeat of Figure BT3. Mohr's circle for (only) the mean components of cyclic bending and torsion. The mean coordinate stress point in this Mohr's circle is the same coordinate point as the alternating coordinate stress point in Figure BT2. A counter-clockwise rotation of (90 - 2θ) degrees on this Mohr's circle for mean stress establishes the values for the mean normal stress and the mean shear stress acting on a plane of maximum range of shear stress.

Remark: When the respective Mohr's circles for *both* mean and alternating stresses are plotted on the same coordinates at time *wt* equals 0 and *wt* equals 180 degrees, the respective maximum ranges of the shear stress and the corresponding ranges of normal stresses are more clearly evident.

CHAPTER 24

Fatigue Remedies and Fatigue Redesign

A single fatigue failure is not necessarily a serious problem. But multiple fatigue failures that occur at the same location on a component and have the same failure appearance are an obvious problem. These failures surely require a remedy. Typically the fatigue failure appearance indicates the type of service loading and the respective magnitudes of the nominal alternating stress and the associated stress concentration factor. Then, because the existing geometry and material specifications are known, the problem is reasonably well-defined and alternative remedies can be evaluated to establish the remedy or combination of remedies that are most advantageous.

It is hard to draw a sharp distinction between fatigue design and fatigue redesign. Fatigue design is a continual process of extrapolating service proven performance to either (i) the design of the next generation of components and machines, (ii) the redesign to overcome current problems with existing components and machines. Accordingly, it is imperative to establish reliable extrapolation methodologies and to understand the analogies that underlie these methodologies. The more reliable this analogy-based extrapolation methodology, the more reliable the sizing calculations and the less need for confirmatory laboratory and simulated-service testing. Unfortunately, this fatigue design process is far from perfect and fatigue failure problems continue to occur. These problems require employing fatigue remedies to fix existing problems, which should subsequently be integrated into the fatigue design process to prevent reoccurrence of such problems. Accordingly fatigue design should be a continual process of improvement. Unfortunately, however, fatigue design is still a learning process whose foundation is usually experience rather than theory.

In my opinion the fatigue literature over the last thirty to forty years has been almost devoid of any data that would actually enhance and advance fatigue design methodologies. Thus it appears that fatigue failure will continue to be the dominant mode of failure for machine components for the foreseeable future.

At present it is virtually impossible to start from scratch in the design of a new machine and its individual components where there is little or no actual service experience with analogous machines and their components, without encountering several serious fatigue problems in its developmental stage. The main problem is that service loads are almost never well-defined for any machine or its components, either with regard to the load-time history associated with certain uses or even with regard to the maximum possible loads at some time in the most extreme yet rational service use. Accordingly, the concept of steady-state operation at the design state of stress, DSS, usually is either fictitious or fantasy. In this situation, confirmatory laboratory testing is clearly inadequate and although simulated-service testing is much better, it is not necessarily reliable.

Because fatigue design is and likely always will be a process of continual improvement, it should also be a process of *a priori* developing alternative generic fixes rather than *a posteriori* specific fixes so that, for any fatigue failure problem that may arise in the future, the most advantageous fix or combination of fixes have already been studied and are readily available and can be easily implemented. Simply said, there should be much more fatigue-design-oriented R&D.

Presently, the methodology for designing against fatigue failure is merely to use fatigue failure problems that occurred in previous designs and were subsequently fixed as empirical fatigue design *don'ts* (don't replicate previous fatigue problems); and to use their fatigue fixes as empirical fatigue design do's. I attempt to enumerate subsequently in this topic some *alternative* design fixes, the do's, given a generic fatigue failure problem.

The concept underlying designing against fatigue failure is that fatigue stress must be less than fatigue strength, where the proper fatigue strength is the notched endurance limit. I assert that if the fatigue stress calculated in the naive factor of safety analysis exceeds the *notched endurance limit*, then fatigue failure will occur. Thus, to fix an existing problem, it is necessary to establish alternative ways to (i) reduce the fatigue stress or (ii) to increase the notched endurance limit, or (iii) accomplish both (i) and (ii).

24.1. ALTERNATIVE WAYS TO REDUCE THE FATIGUE STRESS

The best time to reduce the fatigue stress is in the *redesign stage* when critical dimensions of the problem component can easily be changed. Then the respective sizes of the problem component cross-sections can be better balanced relative to full use of the component material. Ideally, (i) the *ratio* of the nominal alternating stress to the notched endurance limit is uniform throughout the component, so that (ii) material can be *re-allocated* from locations where fatigue failure did not occur to increase the size of the cross-section where fatigue failure actually occurred. This re-allocation process is best carried out using the naive notched endurance limit factor of safety as the computational metric to make the ratio of the nominal alternating stress to the notched endurance limit as uniform as practical. This naive notched endurance limit factor of safety is typically based on a *parametric* value for the nominal alternating stress and fixed estimated value for the notched endurance limit.

Suppose that the dimensions of the *other* cross-sections cannot be changed. Then the required increase on the size of the critical cross-section where fatigue failure actually occurs can still be calculated by performing the respective naive notched endurance limit factors of safety at cross-sections that did not fail to establish a common-sense desired value for this naive notched endurance limit factor of safety at the fatigue critical cross section. It is possible that this calculated change in size merely requires a change in dimension on the component drawing and minor retooling. It is even more likely that the stress concentration at this cross-section can be reduced at the same time to reduce the dimensional changes needed.

Suppose that no dimensions can be changed, than the only other primary remedy on the stress side of the fatigue stress-fatigue strength inequality is to reduce the stress concentration and in turn reduce the value for the associated endurance limit reduction factor.

The expression underlying the beginning part of an efficient methodology in a preliminary design against fatigue failure in which the naive notched endurance limit factor of safety has a value that is as close to being a constant throughout the entire component of member as practical, is:

$$\text{Naive Notched Endurance Limit Factor of Safety} = \frac{\text{Unnotched Endurance Limit}}{K_f * S_{\text{nominal alternating stress}}} = \text{Constant}$$

In turn, the following expression is merely an algebraic re-expression of the naive notched endurance limit factor of safety expression:

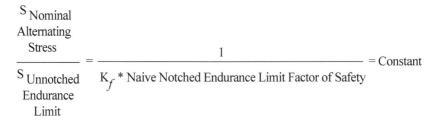

Note that the value of the constant in each of these expressions is critically dependent on the actual stress concentration factor, K_t, but only through its relationship to K_f, the *endurance limit reduction factor*, viz.,

$$K_f = 1 + (K_t-1) * q$$

See the topic entitled "Endurance Limit Notch Sensitivity of Round Mild Steel Specimens", Figure Q2. (Chapter 19)

The most intuitive (and effective) way to understand the magnitude of a stress concentration factor is to understand the mathematical analogy between the curvature of streamline flow in the presence of an obstacle and the magnitude of the stress concentration factor, viz., the sharper the curvature of the streamline, the greater the stress concentration factor. Accordingly, the smoother the streamline, the smaller the stress concentration factor. For example, consider Figure FRR1. The semicircular notch in (a) is an obstacle to streamline flow and all streamlines must change their direction to pass by this obstacle. The streamline curvature is sharpest at the root of the notch and becomes smoother and smoother further away from the notch. In turn, when the semicircular notch in (a) is augmented with two smaller semicircular notches as is (b), these smaller notches cause the streamlines to start to change their direction more gradually, thereby reducing their curvature in the vicinity of the bigger notch and reducing its stress concentration factor.

Remark: The experimental stress analysis technique of photoelasticity has very useful application here in selecting the "optimal" locations and sizes of these so-called stress-relieving notches.

Figure FRR2 is intended to encourage you to visualize streamline flow *and to avoid* geometric shapes that require the streamline flow

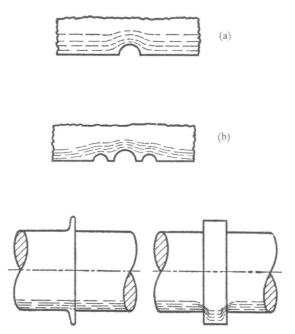

Figure FRR1. Example of the streamline flow analogy for understanding the local intensity of the stress, viz., the magnitude of the stress concentration factor. The large semicircular notch in (a) causes the streamline to curve sharply around the notch, causing a large stress concentration factor. The addition of two smaller notches, one on each side of the large notch, causes the streamlines to start to change their direction further away from the large notch, thereby making the streamlines smoother in the vicinity of the large notch. These smaller notches are typically called *stress-relieving* notches because the stress concentration factor pertaining to the large notch is reduced (and their respective stress concentration factors are smaller than the reduced stress concentration factor).

to change direction abruptly. Question: If the change for (a) to (b) in Figure FRR2 did not solve the fatigue problem, how could you remove more material to further enhance streamline flow?

The simplest way to reduce stress concentration factor for shafts with shoulders is to increase the radius of the fillet. However, Figure FRR3 is intended to suggest intuitively that different blending geometries will have different stress concentration factors. If so, it surely makes sense to examine alternative blending geometries in an attempt to reduce the value for K_t. In particular it seems intuitive to employ the most *streamlined blending shape* that is available for critical applications. Figure FRR4 presents a classical "don't" that keeps reoccurring. It might seem obvious not to superimpose a spot face into a fillet stress concentration, but it still

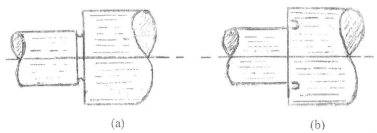

(a)   (b)

Figure FRR2. (a) The way I was taught as a tool and die apprentice to undercut the shoulder on a shaft to permit the grinding wheel on an O. D. grinder to get to the shoulder and even with a little wear at its leading edge but still not obstruct the press fit of a bearing or gear against the shoulder of a shaft. From a streamline flow and a stress concentration perspective it is an extremely poor undercut. In addition, the (smaller) shaft diameter is reduced and the associated nominal stress is increased. (b) A much better way to accomplish the same goal by cutting parallel to the diameter into the shoulder. Then the stress concentration and the nominal stress are both reduced. In turn, with regard to wear of the leading edge of the grinding wheel, extra care can be given to making sure that the external radius on the inner race of the bearing and the size of the chamfer on the bore of the gear will not prevent shoulder contact. Do you think you could further reduce this stress concentration by making a larger, better shaped undercut given that an adequate bearing area (for thrust) could be as small as one-sixteenth wide ring around the O. D.?

commonly occurs. How would you redesign this sand casting to prevent failure while retaining the spot face. Hint: Read the next paragraph.

Sometimes both the nominal stress and the stress concentration factor can be reduced at the same time. This situation generally falls under the heading of *redistribution of nominal stress* and applies more generally to the structural aspects of machines, e.g., housings, frames,

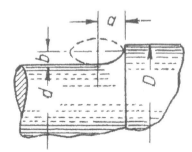

Figure FRR3. Typically, when a shaft has a change in diameter, a fillet is used to blend from one diameter to the other. The larger the fillet employed the smaller the associated stress concentration factor. However a quadrant of an ellipse can reduce the stress concentration factor even further.

brackets, etc. Redistribution of nominal stress is enhanced by conceptualization of *force flow* in much the same manner as the visualization of streamline flow. Then structures are in turn thought of as consisting as a network of linear springs. Accordingly, *the stiffer the member (region), the larger the proportion of the load that it carries.* The goal, however, is to have a number of members (regions) with the same stiffness (spring rate) so that the force is distributed more uniformly. In particular, it is important to make sure that individual members (regions) do not carry what would typically be modeled as "concentrated forces". For example, it is often necessary to employ a gusset to provide rigidity to a portion of a frame or housing. However, several smaller uniformly spaced and appropriately oriented gussets will always be more effective and efficient relative to redistribution of nominal stress, because both the nominal stresses and their associated local stresses are smaller for small individual gussets than a single large gusset.

Another concept that is often useful in the redistribution of nominal stress is that loads are resisted more efficiently by members (regions) in direct tension and in direct compression than by members that resist loads by having to bend and/or twist.

Figure FRR4. The classical superimposition of a stress concentration on a stress concentration. It is the superposition of a spot face on a fillet in this unsophisticated example of a bracket on a commercial mechanical shaker. However, the same effect occurs whenever a stress concentration is inappropriately placed in any region of high stress. For example, drilling a hole near the edge of a member in bending rather than at the neutral axis. Considering this specific example, this cast aluminum bracket could have had its fillet blended into a boss or pad and then the boss or pad could have been spot faced.

Recall also the bolted joint example where the stiffness of the tension nut was reduced by tapering it to reduce its thickness in the region of the first and second engaged threads of the undercut bolt. This same idea is employed in shrink-fit of the hub of a cast iron flywheel on a steel shaft to prevent fretting fatigue failure of the shaft. See Figure FRR5. Suppose in contrast that the press-fit pertains to a gear with a *fixed* uniform width and fretting fatigue failures of the shaft occurred at the start of the press-fit. What remedy would you suggest to eliminate this problem?

Fretting fatigue is discussed briefly under the heading: Alternative Ways to Increase the Endurance Limit.

Figure FRR6 is a photograph of a fatigue crack on the front side of a relay can in a missile. This crack was caused by an arc weld that was located where it was stressed by the harmless oil-canning lateral motion of the side the aluminum sheet metal can. The fatigue problem was solved by simply removing the arc welds. These welds were not needed because each of the can bracket side flanges has five spot welds in a well-designed pattern. This may be a trivial example, but it illustrates the fundamental concept that the addition of so-called redun-

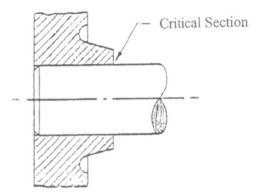

Figure FRR5. A revised geometry intended to redistribute the compressive normal stress of a shrink-fit of the bore of a cast iron flywheel on a steel shaft. The critical cross-section of interest is always at the leading edge of the shrink-fit where fretting failure of the shaft is a critical concern. (Recall that Wöhler had fretting fatigue failures with his press-fit axle specimens.) If the shrink-fit involves too much interference, the cast iron hub may fracture. If the hub is sufficiently thick to resist failure due to its shrink-fit induced static tensile hoop stress, the press-fit superimposes a radial normal stress on the bending stress at the leading edge of the hub. The revised geometry shown employs the relative stiffness principle in an attempt to smooth the transition of the shaft stress from being a state of biaxial compression to becoming more of a nominal bending stress state. What would you recommend if this geometry does not eliminate fretting fatigue failure?

Figure FRR6. A relay can with two redundant arc welds, front and back, that developed fatigue cracks that subsequently propagated into the relay can support bracket. The issue here is that these arc welds resisted the deformations caused by harmless oil canning of these two sides under vibratory loading (despite the stiffening X beads). The fundamental issue here is that *loads can only be developed when there is a resistance to deformation or movement.*

dancy in even a simple structure can cause problems. The best designs are always the ones where all load paths are simple and well-defined, and where the geometry of each member and component is deliberately designed to resist specific loads. Welds and fasteners should be placed only where there is an actual need for them.

There are number of secondary ways to reduce the fatigue stress such as (i) reducing the tensile mean stress by various mechanical methods of producing a counteracting compressive stress, (ii) making sure that the frequency of the cyclic loading does not coincide with a natural frequency of the machine components or structural members, (iii) improving alignment of members in tension, or (iv) perhaps even simply designing for easy maintenance.

24.2. ALTERNATIVE WAYS TO INCREASE THE ENDURANCE LIMIT

Fortunately, examination of the fatigue failure itself typically suggests one or more remedies and perhaps even the most direct and effective remedy or collection of remedies. In some cases the main problem is to select among several viable alternatives.

The naive notched endurance limit factor of safety expression clearly indicates that an increase in the unnotched endurance limit is just as effective as reducing the nominal alternating stress. Moreover, all generic design curves for mild steel that provide an estimate of the unnotched endurance limit are based on the relationship between the unnotched endurance limit and the ultimate tensile strength (or hardness). Thus the first way to increase the factor of safety against fatigue is to select a mild steel with a higher ultimate tensile strength.

Remark: In my experience, one of the *least understood* areas in designing against fatigue failure is the issue of hardness versus hardenability. A competent designer understands that the hardness distribution across the cross-section of bar or rod depends on its size and processing and/or heat-treating. Hardenability is the ability of a heat-treated steel to attain a more uniform hardness distribution across its cross-section. Certain alloy steels can attain a hardness distribution that is (almost) uniform for practical purposes across the entire cross-section, given certain standard stock sizes. These alloy steels are expensive and are often inappropriately selected to fix fatigue failure problems. It is clearly foolish to pay for strength you do not need. I categorically assert that at least ninety percent of selections of alloy steels with superior hardenability that I have encountered in my consulting merely increased the hardness in regions that had nothing to do with the given fatigue failure. AISI 4140 and 4340 are so over-used in fatigue that it defies common sense.

The issue of increasing either size or strength is illustrated in Figure FRR7. In (a) it is clear the stress at the surface exceeds the strength. In (b) the strength at the surface exceeds the stress at the surface because the strength is increased; whereas in (c) the strength at the surface exceeds the stress at the surface because the size is increased. Both (b) and (c) are actually quite poor with regard to using strength efficiently across the entire cross-section given either bending or torsional loads or even combined bending and torsional loads, and especially given a notch.

Remark: Consider an axially-load mild steel member. Suppose that its ultimate tensile strength (Brinell hardness) is perfectly uniform across its entire (stock) cross-section. If the unnotched endurance limit is one-half of the ultimate tensile strength, if $K_f = 2$, and if the naive notched endurance limit factor of safety is equal to two, then the effective utilization of the ultimate tensile strength of

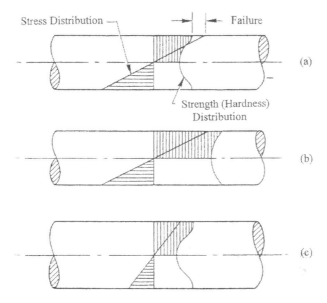

Figure FRR7. Consider the respective stress and strength distributions in (a). In turn, suppose that (b) and (c) represent alternative fixes. In (b) the nominal stress remains the same and a new material with a higher strength is employed to generate an acceptable fatigue factor of safety. In (c) the size is increased locally while maintaining the same material (same strength). What would you recommend?

this mild steel member is twelve and one-half percent. Clearly this utilization is *even worse* for bending and torsion!

The hardness of heat-treated plain carbon steels increases with increasing carbon content. However, the hardness peaks out at Rockwell C hardness levels about eighty to ninety times the percent carbon content, e.g., do not try to heat-treat an AISI 1045 shafting steel to above about 36 to 40 Rockwell C. Always examine the microstructure and, in particular, always temper the member or component to eliminate completely untempered martensite.

Just as on the stress side of the stress-strength inequality there is nominal stress and local stress, on the strength side of this inequality there is stock (bulk) strength and local strength. Processes like induction hardening and flame hardening can increase the local strength of medium carbon steels at the root of notches to a sufficient depth to surround the local stress with its very steep stress gradient as in Figure FRR8 (c). If the carbon content of a low carbon steel is not sufficient to attain the depth of hardness needed by heat-treating, carburizing or nitriding could provide a practical way to overcome the carbon deficiency problem.

The hardness of the local strengthening needed to prevent fatigue failure at the root of the notch can be estimated using the empirical relationship

$$\frac{(K_f \text{ existing})}{(K_f \text{ desired})} = \frac{(\text{existing Ultimate Tensile Strength @ notch root})}{(\text{desired Ultimate Tensile Strength @ notch root})}$$

in which

$$K_f \text{ desired} = \frac{1}{\text{Constant} * \text{Notched Endurance Limit Factor of Safety}}$$

and the *desired* ultimate tensile strength is restated in terms of local hardness using the relationship between hardness and ultimate tensile strength. The required local hardness measurements require sectioning the component or member and as an added benefit, an examination of the microstructure can detect potential problems like decarburization and microstructural anomalies.

Remark One: The *margin of safety* is deliberately made equal in (b) and (c) in an attempt to stimulate you to consider the difference between margin of safety and factor of safety.

Remark Two: The issue of hardness versus hardenability and strength distribution is never important for hollow shafts. The issue then is the economics of producing a hollow shaft. The counter balance to cost is less weight, which may be important in certain applications.

Remark Three: Given that increasing the stock strength of mild steels always cost more money, I recommend never paying for increased strength at locations where fatigue failure is not a problem. Typically this means it is cheapest to solve local problems locally.

Recall that good fatigue design always favors the use of a generous fillets at *internal* corners to reduce the associated stress concentration. Similarly, generous radii should also be used to prevent fatigue crack initiation at sharp external corners and edges even where there is no

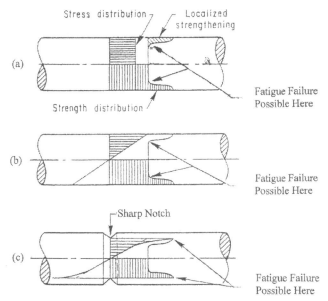

Figure FRR8. Effect of local strengthening on the endurance limit of round mild steel components under alternating loading. (a) Surface hardening decreases very markedly the likelihood that a fatigue crack will initiate along the surface hardened region. Surface hardening also increases local wear resistance. Only a very thin hardened case is required in either of these situations. Deeper case hardening creates an increased tensile stress in the core which might possibly cause fatigue failure to initiate in the core just below the case hardened surface. (b) Same as (a) except the possibility of fatigue failure initiating at the core just below the case hardened surface is perhaps slightly greater. (c) Local surface hardening at the root of the notch is very effect in decreasing the likelihood of fatigue failure initiating at the root of a notch. Overall, the fundamental concept is that local surface hardening solves *local* problems. Finally, returning to (a), I warn that although local surface hardening may also be helpful in resisting corrosion locally, the problem that corrosion is usually ubiquitous and other remedies are more effective.

obvious stress concentration. *External sharp corners and edges are an extremely common initiator of fatigue failures.* Sharp corners and edges, at higher magnification, are actually jagged, almost like saw teeth especially when the direction of machining is orthogonal to edges. In fact, a sharp corner edge may be considered to be a series of notches, each increasing the probability that at least one notch initiates a fatigue crack. Improving the surface finish only produces more notches of more uniform depth. Sharp corners become a very much more serious problem as the hardness of the mild steel components increases. Even chamfers may not be an effective fatigue remedy if one sharp corner is actually replaced by two moderately sharp edges. Thus the effectiveness of chamfering as a fatigue remedy is never as good as when the sharp corner or edge is carefully broken by hand by filing or when a fillet is machined along the sharp corner or edge. The most cost-effective fatigue remedy may be shot-peening sharp corners and edges in certain situations.

Remark: Shot-peening was used to eliminate fatigue failures of the stamped metal spider of the vintage automotive fan blade with solid blades riveted to the spider. Ironically this fatigue problem reappeared when Ford Motor Company introduced the flex-fan. Then, before its fretting fatigue problem became well-known, Chevrolet adopted the flex-fan in its bestselling pickup truck.

Sharp edges caused by shearing sheet steel are especially detrimental to the endurance limit. The problem is compounded when subsequent forming involves direct stretching or bending such that the tear side of the shear edge is stretched. In cases where a stamped mild steel component is, or may be, stressed in the direction along a sheared edge, subsequent shot-peening can be used in fatigue-critical regions. When a flange is a potential fatigue initiation site, a hemmed flange, made by folding the flange back on to itself, is a very effective fatigue remedy.

In my experience, the major fatigue problem with sheet metal flanges is local buckling of the flange at the compression side of a bending operation of forming. I once worked on a case where the bestselling car of an automotive manufacturer that had two bucket seats, each with two seat track brackets whose stamping always had a flange that buckled to form a hemmed flange *in reverse,* viz., the hem-line seam was visible looking at the outside edge of the flange. This seam opened and closed with each forward and backward movement of the seat. A few months after stating in response to a discovery document that the manufacturer knew of no serious problem with these brackets, the company issued a recall that involved replacing all four seat-track brackets in the bestselling car and a companion car. Although the buckling of this seat-track bracket flange was an extreme case for a product that unfortunately got into service and then caused a serious problem. I have seen local buckling of sheet metal flanges to some degree on almost all of the stampings that I have examined.

The king of all sharp corners and edges is the *feather edge*. Figure FRR9 is a photograph of a forged aluminum hydraulic cylinder that was used in older aircraft that had its stroke extended a short distance by modifying the shape or its inlet port geometry to permit drilling a hole from the end of the cylinder at a upward angle to intersect the *shortened* hole in the inlet port. The hole in the cylinder bore thus became an ellipse instead of the former round hole. Although the stress concentration was increased by the presence of an ellipital hole, the real fatigue problem was the feather edge at the forward end of the elliptical hole. This modified cylinder was subsequently used in a mechanism to open a set of flaps so that the front wheels of a four engine aircraft could be lowered. The cylinder failed to actuate because of the fatigue crack in the photograph that originated at the feather edge and propagated through the cyclinder wall to cause severe leakage. When I asked if the extended stroke was actually needed for the incident aircraft I was told it was not. Then I suggested that the inlet port hole merely be drilled all the way through the cylinder wall (and not drilling the intersecting hole). This fix solved the fatigue problem.

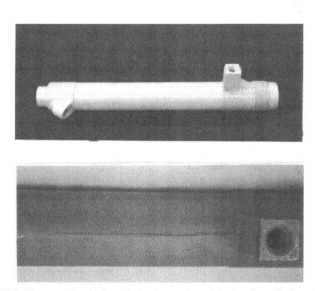

Figure FRR9. Photograph of a forged aluminum aircraft hydraulic cylinder whose original inlet boss was modified to increase the stroke slightly. This modification caused a feather edge at the inside surface of the cylinder, but I have no information regarding the actual performance of this modified cylinder in its previous application. This modified cylinder was unfortunately adopted for use in another aircraft resulting in the fatigue failure that I encountered in my consulting. In this application a fatigue crack initiated at the feather edge of an elliptical hole at the inside surface of the cylinder and propagated longitudinally along and through the cylinder wall until excessive leakage caused the cylinder to fail to actuate. As discussed in the text, the solution was to eliminate the feather edge. (This failure is also important as a reminder to know the complete service history of any existing component being considered for use in a new application.)

Remark: A feather edge in certain locations along its length can be sharper than a surgical scalpel. If it is alternately loaded in tension at any of these locations, a fatigue crack is *almost certain* to initiate.

Fatigue failure is almost certain to occur in all cases involving *continuous* surface damage. The only issue is the length of the fatigue life and it may be measured in time as much as in cycles. Thus unless effective preventative measures are taken, fatigue failure of mild steel can be considered inevitable in design situations involving fretting, scoring, or corrosion. See Figure FRR10.

The common remedy for all surface-damage-related fatigue problems is local surface hardening. I have never seen a wear or scoring surface damage problem that was not eliminated by local hardening

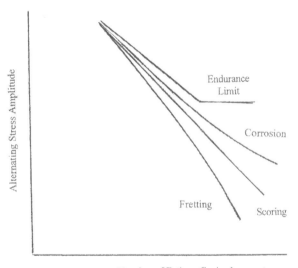

Figure FRR10. Schematic of effect of continual surface damage on the fatigue strength of mild steels. (i) There is no endurance limit. (ii) The fatigue life given continual fretting or scoring is typically more a function of the number of cycles rather than duration. My conjectured "typical s_a-N curve" for fretting has a reverse curvature because it takes time to begin, but once fretting debris is formed, fretting continually increases in intensity as the number of fatigue cycles increases. (iii) The surface damage due to scoring usually increases more linearly with the number of imposed fatigue cycles than does fretting. (iv) The fatigue life given continual corrosion is typically more a function of time than for fretting or scoring, particularly given for relatively mild corrosive environments. Thus a typical s_a-N curve for mild corrosion still has a small forward curvature.

and its associated compressive residual stress. If local surface hardening does not solve a fretting fatigue problem, it will surely lengthen the fatigue life. In some cases just surface compressive stresses may suffice to solve the fatigue problem. If so, surface rolling and shot-peening can be recommended as a remedy.

Fretting is caused by continual *minute* rubbing of two surfaces together under steady pressure, e.g., as in a shrink-fit. The surface finish asperities rub across one another resulting in local welding. If fracture occurs in the softer material away from the weld interface, it causes wear debris that oxidizes and increases in both volume and abrasiveness, digging pits in the softer surface. This debris cannot escape so that the pits just get more numerous and bigger and bigger until a fatigue crack initiates at one or more pits. Then failure is inevitable when fatigue loading is continual. The fundamental remedy is to prevent local welding. Minute motions is very difficult to eliminate or even control. Local welding is diminished when the two materials are dissimilar. Local welding is diminished or perhaps eliminated by an appropriate lubricant that acts as anti-flux and cannot be squeezed out from between the two surfaces. Putting these two facts together suggests that a soft plating such as zinc, cadmium, copper, or nickel is perhaps the best remedy. Sometimes, however, it is possible to redesign so that rubbing surfaces are physically separated by shims, inserts, or gaskets made of rubber, plastics, or reinforced resins.

In scoring, as opposed to fretting, wear debris can be flushed from between the rubbing surfaces by the motion of the lubricant that is usually recycled after passing through a filter. The incompatibility of the material pairs with regard to local welding and the ability and tenacity of the lubricant to wet the surface are important factors in reducing or eliminating scoring. Moreover, in contrast to fretting, the depth of the surface hardening can be only a few thousandths of an inch deep to be very effective in reducing or eliminating scoring. In automotive engines the crankshaft journals are made much harder than the journal bearing, so that when bearing failure eventually occurs, the bearings are replaced rather than the crankshaft. Metals that develop strong, tough oxide layers resist scoring. Given mild steels, the anti-scoring equivalent to this oxide layer can be attained by surface coatings and platings.

Remark One: One of the fundamental rules in machine design is to make replacement of all bearings as simple and easy as possible, regardless of the type of bearing. I have seen both shaft fatigue failures, shaft scoring failures, and excessive plastic deformation failures all occur as a direct consequence of a prior bearing failure. Typically the bearing failures that caused these subsequent failures occurred because the bearings were not replaced or repaired in a timely manner.

All corrosion remedies are also corrosion-fatigue remedies. The resistance of mild steels to corrosion fatigue does not depend on their ultimate tensile strengths. In addition to local surface hardening, only surface treatments such as spray metallizing and various coatings are effective in prolonging fatigue life in the continual presence of a corrosive environment. In fact, it is usually recommended to apply surface treatments and coatings in addition to surface-hardening.

Remark Two. The ability of a metal to develop an oxide layer that resists further corrosion is no guarantee that corrosion fatigue will not occur. The first replacement hip joints employed 316 stainless steel for its *well-known* resistance to corrosion. However, the minute relative movement of the 316 stainless steel stern relative to the femur bone continually wore off the oxide layer and subsequent corrosion fatigue caused the average life of these hip joint stems to be about 36 months.

The effect of surface finish is always discussed· in machine design textbooks. However, as I mentioned when discussing notch sensitivity, almost all notches as tested as machined without additional polishing. Accordingly the effect of machining is typically built-in to the endurance limit notch sensitivity factor. No one ever polishes a hole (the correct term is lapping), and if it did happen there would be a feather edge at it's beginning and ending, unless the hole were chamfered; and I have never seen a polished chamfer, counter-sink, or spotface. We cannot even EDM a very sharp uniform circumferential V-notch, so how can we polish it. I admit that I once saw a paper in which the fatigue investigator used a tiny diameter copper wire dipped in an alumina slurry to polish a not very sharp V-notch. I remember thinking then that he was a nut! In fact, after reflection, I still do. I have never heard of anyone who has ever polished a keyway or a key seat. Fatigue investigators polish round fatigue specimens—and as experience clearly shows that they do not do that very well. In fact, to this day and likely forever, no fatigue specimen polishing technique has ever passed a round robin test program.

Final Remark: I have bragged for over fifty years that I know an exception in the fatigue literature to every quanitative or qualitative assertion about fatigue behavior found in a machine design textbook. I now brag that about forty years ago that I machined circumferential V-notches in round mild steel specimens such that the fatigue strength of my notched laboratory fatigue specimens was ***greater*** than the fatigue strength of my unnotched laboratory fatigue specimens. This trick was accomplished by controlling the rake angle of my tool bit so that I put so much cold-work at the root of circumferential groove that when the fatigue crack initiated, it was not at the root of the V-notch, but on the flank of the V-notch approximately half-way up to the specimen diametral surface. It then propagated all the way around the bulb-like cold work region to a point almost directly below the notch root before turning and propagating more or less directly toward the center of the specimen and ultimately to abrupt failure.

CHAPTER 25

Bolted Joints

BOLTED *joints should always be pre-tensioned so that the bolt can exhibit an increased endurance limit relative to no pre-tensioning.* To understand this assertion, I will first explain the mechanical aspects of a pre-tensioned bolted joint. Figure BJ1(a) depicts my two exemplar bolts (machine screw with a nut) being used to clamp a steel cylinder head to a cast iron pressure vessel. (This sketch shows only two of the 2∗N or 2∗(N)+1 bolts equally-spaced around the so-called *bolt circle*.) Note also that the thickness of the cylinder head is different than the thickness of the cylinder flange. (This is an important issue.) Figure BJ1(b) also depicts two exemplar bolts clamping two halves of a solid coupling together. If the primary purpose of this solid coupling is to transmit torque from one rotating shaft to the next rotating shaft, one or more of the respective 2∗N or 2∗(N)+1 bolt shanks must experience both a compressive bearing stress and a transverse shear stress. However a better design would have two hardened dowel pins located 180 degrees apart on the bolt circle that would resist the torque that one side imposes on the other side. The bolts would then be loaded only in tension and the torsion associated with friction resisting their respective pre-tensioning tightening processes.

Now looking down directly at the cylinder cap in Figure BJ1(a), imagine a typical pie-shaped segment of the cylinder cap containing a centered bolt and the cylinder flange below. Figure BJ2(a) depicts the presumed linear tensile force-elongation relationship for this bolt, so that it can be modeled as a linear spring in subsequent analysis. Figure BJ2(b) displays the corresponding presumed linear compressive force-compressive deformation relationship for the localized regions of the clamped cylinder head and the cylinder flange below around the respec-

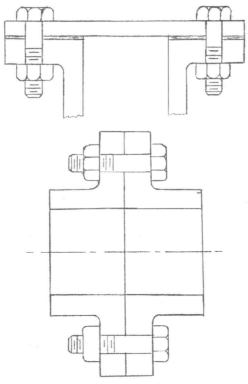

Figure BJ1. (a) A bolted cylinder head, and (b) a bolted solid coupling. These two exemplars were selected because the associated statically-equivalent bolt forces and statically-equivalent clamped-parts forces are relatively intuitive.

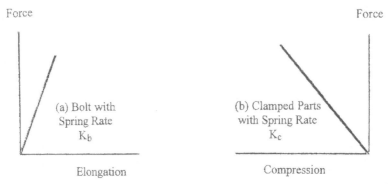

Figure BJ2. The presumed linear relationship between the statically-equivalent tensile bolt force and the nominal increase in bolt length (a) and the presumed linear relationship between the statically-equivalent compressive clamping force and the nominal decrease in the thickness of the clamped parts (b).

tive bolts. These localized regions are also modeled as linear springs that are compressed by their centered bolts.

Figure BJ3 presents a schematic of the statically equivalent forces and their associated nominal deformations when the exemplar centered bolt is pre-tensioned in accordance with the calculations in Appendix B. During the pre-tensioning process, the exemplar centered bolt elongates whereas the associated portion of the clamped parts are compressed, each acting according to their respective idealized linear force deformation model. The exemplar pre-tensioned bolted joint pie-shaped segment is in static equilibrium because no external loads act on this free body segment. Thus the respective statically-equivalent bolt and clamping forces must be equal and opposite. However if some statically-equivalent (external) applied force were somehow applied to the bolted joint, the bolt would be further elongated by exactly the same amount that the compression of the clamped parts would be reduced. Moreover, the sum of the resulting increase in the statically-equivalent tensile bolt force and the resulting decrease in the statically-equivalent clamped-parts compressive force is equal to the (external) applied force. See Figure BJ4.

The increase in the statically-equivalent bolt force is equal to $[(K_b)/(K_b + K_c)]$ times the statically-equivalent (external) applied force, whereas the decrease in the statically-equivalent clamping force is equal to $[(K_c)/(K_b + K_c)]$ times the statically-equivalent (external) applied force. Then the bolt alternating stress is simply one-half of its increase and the mean stress is the initial tightening (static) stress plus one-half of its increase. See Appendices A and B.

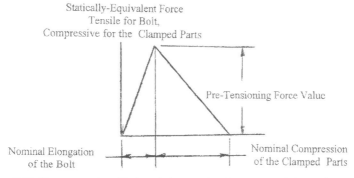

Figure BJ3. The pre-tensioned bolt and the associated compressed clamped parts. Both experience the same statically-equivalent force, tensile for the bolt and compressive for the clamped parts, equal in magnitude and opposite in direction, to satisfy static equilibrium.

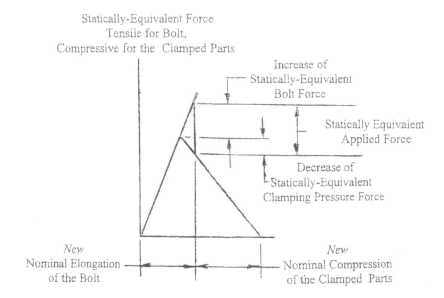

Figure BJ4. The pre-tensioned bolted joint under the action of a statically-equivalent (external) applied force. Clearly this statically-equivalent applied force, if it were sufficiently large, would open the joint so that all of this statically-equivalent applied force would then have to be resisted by just the bolt. This of course cancels the benefit of pre-tensioning. However, it is also clear that for all smaller statically equivalent applied forces, the statically-equivalent clamping force is reduced and the statically-equivalent bolt force is increased. Moreover, because the bolted joint does not open for this statically equivalent (external) applied force, the increase in the nominal length of the bolt must be exactly equal to the decrease in the nominal thickness of the clamped parts. Accordingly, the nominal spring rates of the bolt and the clamped parts can be used to compute the proportion of the statically-equivalent (external) applied force that increases the bolt statically-equivalent force and the corresponding proportion of the statically-equivalent (external) applied force that reduces the statically-equivalent clamped parts force.

25.1. APPENDIX A

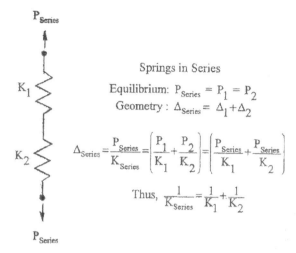

Springs in Series

Equilibrium: $P_{Series} = P_1 = P_2$
Geometry: $\Delta_{Series} = \Delta_1 + \Delta_2$

$$\Delta_{Series} = \frac{P_{Series}}{K_{Series}} = \left(\frac{P_1}{K_1} + \frac{P_2}{K_2}\right) = \left(\frac{P_{Series}}{K_1} + \frac{P_{Series}}{K_2}\right)$$

Thus, $\dfrac{1}{K_{Series}} = \dfrac{1}{K_1} + \dfrac{1}{K_2}$

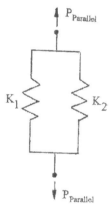

Springs in Parallel

Geometry: $\Delta_{Parallel} = \Delta_1 = \Delta_2$
Equilibrium: $P_{Parallel} = P_1 + P_2$

$$P_{Parallel} = P_1 + P_2 = K_{Parallel}\Delta_{Parallel} = K_1\Delta_1 + K_2\Delta_2$$

Thus, $K_{parallel} = K_1 + K_2$

25.2. APPENDIX B

The bolt spring rate is calculated by treating the respective shank and threaded portions of the bolt along its grip length as spring and series. viz.,

$$\frac{1}{K_{Series}} = \frac{1}{K_{Shank}} + \frac{1}{K_{Threaded}}$$

$$\frac{1}{K_{Series}} = \frac{L_{Shank}}{EA_{Shank}} + \frac{L_{Threaded}}{EA_{Threaded}}$$

Thus

$$K_{Bolt} = \frac{L_{Shank} A_{Threaded} + L_{Threaded} A_{Shank}}{EA_{Shank} A_{Threaded}}$$

The spring rate calculation for the clamped parts will be treated later.

Bolts are different than other machine components in that it is the proof stress that limits the maximum value of the nominal normal stress, viz., the proof stress is the *surrogate* for the yield strength in bolt tensioning calculations. Moreover, the associated bolt nominal normal stress in bolt tightening calculations also employ a committee-based stress area that is widely tabulated. I recommend that you never design using a nominal normal stress greater than 0.90 times the tabulated proof stress. For example, consider a 1/2 - 13 UNC, grade 8 machine screw. Its proof stress is 120,000 psi and its stress area is 0.1419. Thus, tensioning the machine screw to 0.90 times its proof stress requires a nominal normal initial tension force equal to

$$120,000 \times 0.90 \times 0.1419 = 15325.2 \text{ pounds}.$$

The torque (in inch pounds) required to achieve this initial tension is computed using the expression $T = CDF_i$, viz.,

Initial Tightening Torque = Torque Coefficient x Bolt Diameter x Initial Tension Force.

The value of the so-called torque coefficient is typically taken as 0.20; but can vary by about minus and plus 50 percent. I am much more confident using the "turn of the bolt" method of tightening, where you calculate the nominal tensile elongation of the bolt for the given grip length and using the known pitch of the thread, compute the rotation amount turn of the bolt required for the given grip length, starting with a very snug "hand tight" situation.

The next issue is never to exceed the proof stress to prevent potential yielding when imposing a statically-equivalent (external) applied force. This means the statically equivalent nominal tensile bolt force can increase by 0.10 times the proof stress times the stress area, i.e., by 120,000 times 0.1 times 0.1419, by 1702.8 pounds to just be equal to the associated "proof force". Next suppose that the gasket factor $[(K_b)/(K_b+K_c)]$ is equal to 0.08 and the imposed statically-equivalent (external) applied force is 10000 pounds. Then the pseudo factor of safety is 1702.8 divided by $[(0.08)(10000)]$ equals 2.185. There is also a *pseudo* factor of safety at the joint opening. This is the initial tensioning force, 15325.2 pounds, divided by (1 - 0.08) times 10000, equals 1. 6658. Next I turn to the calculation of K_c.

The concept of a pressure cone being generated under the bolt head and under the corresponding nut upon tightening was evidently initiated by Rotscher. Figure BJB1(a) displays Rotscher's conjectured 45 degrees pressure cone in his 1927 book entitled, "Maschineneelemente". However research following World War II indicated the actual pressure cone angle is less than half of his conjectured 45 degree angle. I recommend an angle of about 15 degrees for relatively thin individual clamped parts up to perhaps 20 degrees for thicker parts.

Remark One: My axial-load fatigue tests on lap-joints with one-eighth inch thick composite materials bolted to hardened steel plates indicate a *negative* angle based on the respective fretting outlines on the faying surface of the composite specimen.

Remark Two: All machine design textbooks in this country used Rotscher's 45 degree pressure cone angle until I wrote an article in Machine Design magazine with the generic pressure cone spring rate expression below. Since then it seems that many if not most authors have opted to reduce 45 degrees to only 30 degrees for *both* upper and lower pressure cones. Thus these authors continue to employ Rotscher's illogical notion that the pressure cone can (must actually) transverse the faying surface even when the respective materials differ. This is absolutely absurd.

Figure BJB1(b) displays a sketch of my pressure cone with an arbitrary angle ϕ. This pressure cone can be viewed as a series of washers of thickness (dx), each with a spring rate (dx)/AE, where A is the washer area $(\pi)[(D(T,\phi)^2-d^2)/4$ where t varies from zero to T, and E is the elastic modulus. The resulting expression is

$$K_c = \frac{\pi E d \tan\phi}{\log_e\left[\dfrac{D(2T\tan\phi+D) + d(2T\tan\phi-d)}{D(2T\tan\phi+D) - d(2T\tan\phi+d)}\right]}$$

This expression is *properly* used to compute both the K_c value for the first pressure cone and its resulting diameter at the faying surface. Then the angle of the second pressure cone is computed to make the respective diameters of the pressure cone contact areas exactly equal at the faying surface. See Figure BJB2.

Case One: Both Clamped Member Are Steel. Start with the thinner member if one member is thicker than the other. Select a pressure cone angle between 15 and 20 degrees and compute the footprint diameter on the faying surface and the corresponding value for K_c. Then, given this footprint diameter, compute the pressure cone angle for the thicker member.

Case Two: Only One Clamped Member Is Steel. Same as Case One except always start with the steel clamped member (or the clamped member with the higher modulus of elasticity).

Gasket Analysis: The gasket in Figure BJ1(a) is also treated as a spring in series with the two clamped members. However, its pressure cone angle is taken at zero degrees. If this gasket is soft (has a low spring rate), then its spring value will dominate the spring rate for the entire collection of clamped parts. Consequently K_b will be much larger than K_c and $[K_b/(K_b+K_c)]$ will approach one rather than zero. Accordingly, the vast majority of the statically-equivalent (external) applied force will be added to the pre-tensioning force acting on the bolt. This situation presents a serious design problem for the bolt whether the actual mode of failure is either yielding or fatigue. In designing against yielding, it is only necessary to keep the maximum value for the nominal bolt tensile stress less than the proof stress (as previously explained). In

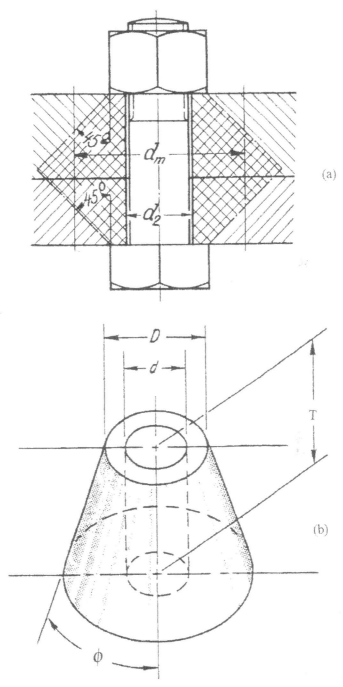

Figure BJB1. Rötscher's 45 degrees pressure cone (a) and my generic pressure cone (b).

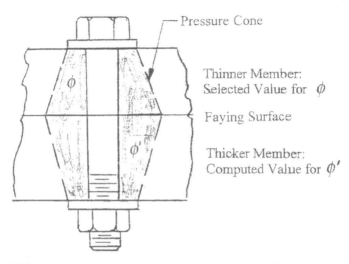

Figure BJB2. My generic pressure cones meet and are the same diameter at the faying surface. This is a straight forward procedure using my generic clamping pressure cones.

designing against fatigue failure, it is typically necessary to use a *hard* gasket with high (higher) spring rate, e.g., a metal gasket and a soft bolt with a low (lower) spring rate, e.g., an aircraft-type bolt, so that the resulting ratio (gasket factor) $[K_b/(K_b+K_c)]$ is ideally 0.10 or less.

Remark: If fatigue becomes a serious problem then the most common remedy is to attempt to make the local force distribution among the threads more uniform. The force distribution among the threads is non-uniform in part because the bolt threads are stretched when the bolt is tightened, whereas the nut threads are compressed. Overall, the first thread typically resists about 40% of the total bolt force, the second thread typically resists about 35% of the total bolt force, and the third thread typically resists about 20% of the total bolt force. (Thus it is clear that the nut need not have more than about four threads.) If possible, use a stud in a blind hole in preference to a bolted joint because both the threads of the stud and the blind hole are stretched in tension, thereby producing a more uniform thread force distribution. Using an aircraft-type bolt with a smaller ("softer") spring rate will reduce the proportion of the (external) applied bolt load resisted by the bolt. Using a tension nut is akin to using a stud because both the threads in a bolt and its tension nut, Figure BJB3, are stretched in tension when the bolt is tightened. The local thread force distribution in the tension nut is improved further by tapering its end as illustrated in Figure BJB3. If corrosion is not a serious potential problem, using an aluminum or brass tension nut will provide an even more uniform local thread force distribution because their modulus of elasticity is lower than that for steel.

25.3. THE COUPLING RE-VISITED

A precision version of the coupling depicted by the sketch in Figure BJ1 (b) is used by riggers to transfer rotary motion to a cantilevered shaft which imposes a rather large statically-equivalent bending moment on the coupling. Using classical rigid-body-based arguments, it is typically alleged that the respective coupling plates display a tendency to separate, maintaining contact only at their bottom edge or at the extreme bottom of the bolt circle. I think this classical analysis is credible only for "finger tight" bolts. Rather, I think the rigid body analysis should pertain to transverse rotation about the centerline of the coupling. Then the bolts above the centerline would be stretched further while the corresponding cramping pressure would be reduced; whereas, in contrast, the clamping pressure would be increased below the centerline while the corresponding bolt elongation would be re-

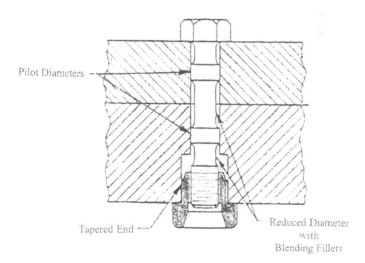

Figure BJB3. Bolted joint with an aircraft-type bolt with pilot diameters and with a tension nut.

duced. Figure BJB4 depicts both actions on the same plot. It is clear that the statically-equivalent nominal *mean* tensile bolt force is equal to pre-tensioning statically-equivalent nominal tensile bolt force. It is also clear the statically-equivalent nominal *range of tensile bolt force is doubled*. The methodology permits the computing the individual bolt force increase and decrease as the coupling rotates. This is accomplished by moving the top bolt starting position for zero to three hundred sixty degrees while computing all the bolt force increase and decrease so that the moment is invariant.

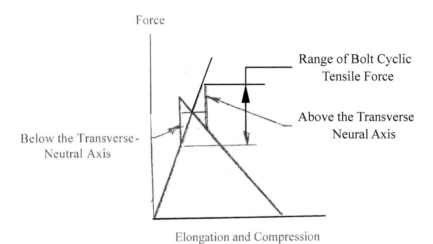

Figure BJB4. Rotating coupling with uniformly pre-tensioned bolts. The transverse neutral axis pertaining to the applied bending moment is stationary. The bolts above the neutral axis experience an increased tensile force, whereas the bolts below the neutral axis experience a reduced tensile force.

Index

1100-F aluminum, 61
316 stainless steel, 168
A-Basis Statistical tolerance limit, 15, 74
abrupt failure, 48, 59, 60, 67, 169
abrupt fracture, 3, 12, 13, 41, 48, 49, 50, 53, 55, 60, 65, 133
aircraft hydraulic cylinder, 165
AISI 1020 steel, 60
ALCOA, 43
alternating normal stress, 4, 5, 27, 35, 81, 91-93, 122, 134, 137
alternating normal stress cycle, 4
alternating stress, 2, 5, 13, 15, 17, 30, 47, 50, 83, 87, 96, 107, 108, 122, 147, 150, 151, 153, 154, 160, 173
alternating stress amplitude, 5, 17, 122
alternating torsional stress, 109, 110
aluminum measuring spoon, 58
amplitude, 5, 17, 18, 24, 26, 47, 48, 78, 81-83, 91-93, 104, 106, 107, 122
axial normal stress, 38, 39, 110, 137
axial-load cycle, 31
axial-load endurance limit, 79, 97, 103, 116, 120, 127-130, 135, 137-139, 145
axial-load fatigue test, 3, 38, 104, 121, 123, 134, 138, 177

batch-to-batch effect, 44, 46
Battelle Memorial Institute, 87, 88
B-Basis Statistical tolerance limit, 15, 17, 18, 74

bearing load, 82
bending failure, 51, 63
bending stress, 2, 48, 65, 85, 115, 123, 158
bicycle handle bar, 54
block-by-block, 82
blunts, 39
bolt circle, 171, 181
bolt spring rate, 176
bolted joints, 171
Box-Cox test, 17
Brinell hardness, 160
brittle failures, 59, 133
"Brown, R. M.", 128

cantilever fatigue specimen, 40
cast steel axle, 1, 3
central bursting, 64
chamfer, 156, 163, 168
circumferential groove, 104, 169
circumferential notch, 103, 119, 121, 142
clamping force, 172-174
cleanliness, 74
collar, 11-13
combined stress, 105-107
competing modes of failure, 74
compressive bearing stress, 171
compressive deformation, 171
compressive force, 171, 173
constant stress, 137, 138
constant stress geometry, 137, 138

183

constrained deformation, 69
constrained yielding, 96, 117, 130, 131, 139, 142
corners, 4, 56, 65, 78, 162, 163, 164
corrosion failures, 47, 58
corrosion fatigue, 3, 168
counter-sink, 168
cramping pressure, 181
crank-driven fatigue test, 33, 79, 137
critical cross-section, 144, 146, 147, 153, 158
critical length, 41, 65
critical stress element, 143, 146, 147
cross-slip, 39, 40
cumulative damage methodology, 82
cup-and-cone failure, 61, 64, 71
cyclic invariant, 102
cyclic loading, 29, 159
cyclic normal stress, 23, 24 39, 128, 129
cyclic softening, 35, 36, 38, 96, 105
cyclic softening effects, 96
cyclic stress, 3, 26, 30, 39, 78, 87, 107, 127, 128, 137, 148
cyclic yield strength, 27, 30, 38, 104, 116, 128-130, 132, 134, 135, 137, 145
cylinder flange, 171
cylinder head, 171, 172
cylindrical plexiglas rod, 63
cylindrical specimen, 3, 79, 114

data based endurance limit, 27, 120
decarburization, 162
deflection-controlled fatigue test,
deformation fatigue, 105
design sizing analyses, 44
design state of stress (DSS), 44

die-cast bracket, 63
dishing, 48, 55
dogbone Plexiglas, 65
dogbone tension test, 62
dual-cantilever rotating bending machine, 3
ductile torsional failure, 59
ductile failures, 59
dwell period, 55
dynamic load, 23

dynamic normal cyclic stress, 23
dynamic normal stress, 23

e_a-N curve, 104
edge effect, 96
effect of hollow specimens, 99
effect of mean stress, 26, 79, 97, 103, 105, 109, 116, 127, 138, 139, 141, 145
effect of shape, 99
effect of surface finish, 77, 98, 168
elasticity, 154
electric absorption dynamometer, 113
endurance limit, 1, 4, 6, 15-17, 19-21, 23-27, 34, 35, 37, 43, 44, 46, 56, 75, 77-79, 83, 85, 91, 95-100, 103-106, 109, 113, 115-117, 119-121, 123, 125, 127-130, 132, 134, 135, 137-139, 141-143, 145, 146, 148, 152-154, 158-160, 162-166, 168, 171
engine mount, 57
Ewing and Humfrey, 39, 40
eyebolt, 55

factor of safety, 6, 27, 77, 125, 143-149, 152-154, 160-162, 177
fail-safe design, 82, 86
failure locus plot, 144
failure state of stress (FSS), 143, 146
faired s_a-N curve, 43, 97
fatigue crack growth rate, 72, 82
fatigue crack initiation, 6, 39, 47, 162
fatigue cracks, 3, 13, 39, 41, 49-51, 55, 82, 133, 159
fatigue cycles, 1, 5, 6, 19, 20, 96, 99, 122, 166
fatigue design, 38
fatigue ductility coefficient, 102
fatigue effects, 75, 86, 98
fatigue factor of safety, 143, 147, 161
fatigue failures, 2, 47, 55, 77, 78, 92, 98, 102, 149, 151, 158, 163, 164, 167
fatigue life, 1, 4-6, 16, 18, 74, 78, 81-83, 86, 88, 89, 96-99, 101, 103, 119-122, 166-168
fatigue redesign, 151
fatigue remedies, 151, 168
fatigue strength, 6, 15-18, 20, 48, 49, 83,

96, 102, 104, 120, 152, 153, 166, 169
fatigue strength coefficient, 102
fatigue striations, 41, 73
fatigue test machine, 1, 2, 25, 33, 34, 37, 78, 79, 85, 86, 88, 89, 105, 113, 114, 119, 134, 135, 137
faying surface, 177, 178, 180
feather edge, 78, 99, 164-166, 168
fillet radius, 4
Findley, 110, 111
finite fatigue life, 1, 6, 121
finite life, 1, 6, 16-18, 33, 95-98, 100-105, 120, 121, 124, 148
finite-life fatigue tests, 101
fracture, 3, 7, 8, 11-15, 41, 48-51, 53, 55, 60, 61, 65, 67, 69, 93, 102, 114, 133, 158, 167
fretting failure, 158
fretting fatigue, 3, 13, 51, 158, 164, 167

gas caps, 58
gasket factor, 177, 180
generalized fatigue models, 109, 122
generic design curve, 134, 138
generic factor of safety, 143
Glocker, 29
Goodman diagram, 4, 23-27, 127, 128
  modified Goodman diagram, 4, 23-27, 127, 128
  modified-modified Goodman diagram, 26, 27
Gunn model, 130

"Haigh, B. P.", 14
Haigh-Soderberg coordinates, 27, 127, 128
hardenability, 76, 160, 162
hardness, 76, 123, 160-163
Hertz stress, 82
hollow hardened steel, 53
hollow specimen effect, 95, 96, 99, 100, 141
homoscedastic normal strength, 16
hysteresis loops, 33, 34, 38, 101, 131

infinite fatigue life, 1
infinite life, 1, 6, 16, 148,
initiate, 3, 12, 39, 48, 49, 51, 53-56, 60, 61, 63, 64, 78, 98, 133, 163, 165-167, 169, 177
interior grains, 29
internal defect, 64

Kc, 173, 177, 178, 180
Kf, 77, 98, 119-121, 123, 125, 130, 135, 142, 154, 160, 162
knee, 1,16, 17, 33, 35, 93, 95-99, 121-123
knee location, 33, 35, 95-99
Kuhn and Hardrath, 120, 123

Lea and Budgen, 113, 115, 116
least-squares analysis, 98
load line, 144-148
load paths, 159
load-controlled fatigue test, 32, 88, 105, 137,
load-controlled test machine, 88
local strengthening, 162, 163
local surface hardening, 163, 166, 167, 168
lubricant, 167
Luder's bands, 62, 110, 131
Ludwig and Scheu, 103, 133

machine component geometry, 3
machine design, 3, 26, 27, 33, 35, 38, 76-78, 82, 85, 89, 92, 93, 95, 103, 105, 110, 122, 128, 129, 134, 137, 141, 167-169, 177
machine effect, 79
margin of safety, 162
maximum value of cyclic stress, 3
mean coordinate stress point, 108, 150
mean stress, 17, 23, 24, 26, 27, 30, 44, 79, 85, 93, 95, 97, 103, 105-110, 116, 117, 127-130, 132, 134, 136, 138, 139, 141, 143, 145-147, 150, 159 173
mean stress diagrams, 44, 79, 105, 143
mean stress effect, 79, 138, 139, 141
mean stress lines, 127
median endurance limit, 6, 17, 19, 20, 43, 46, 75, 91, 97
median-bias-corrected estimates, 17
medians, 6, 121

Method Two Design Factor of Safety, 147
micro-cracks, 65
micrograph, 60, 61, 67, 68-74, 102, 133
micro-plastic, 33-35, 38
microstructural anomalies, 162
mild steel shafts, 59, 113
mild steel specimens, 25, 26, 33, 34, 77, 79, 85, 95, 97, 101, 103, 105, 106, 113, 117, 119, 121, 122, 127-130, 133-137, 139, 141, 142, 145, 148, 154, 169
ML analysis, 16, 18, 20
mode of failure, 21, 47, 67, 68, 78, 91, 92, 143, 144, 152, 178
mode of loading, 17
Mohr's circles, 107, 108, 116, 143, 147, 148, 150,
"Moore, R. R", 33-36, 81, 105, 114, 119, 141

NACA, 87, 88
NACA Report 1190, 87, 88
Naïve, 87, 88
naive factor of safety, 6, 27, 143, 144, 146, 152
naive fatigue effect, 75
naive notched endurance limit factor of safety, 125, 153, 154, 160
NASA, 87, 91, 123
negative q, 121
Neuber, 119
Neuber's rule, 103
Nf, 103
Nishihara, 83, 96, 109, 110
Nishijima, 43
normal strain, 34, 63, 131
normal stress, 2, 4, 5, 23, 24, 27, 34, 35, 38, 39, 42, 55, 77, 81, 91-93, 95, 108, 109, 114-116, 120, 122, 129, 130, 134, 137, 143, 150, 158, 176
notch root, 60, 98, 162, 169
notch sensitivity, 77, 79, 103, 104, 119, 120, 123, 138, 142, 154, 168
notch strengthening effect, 92, 96, 104, 119, 132, 138, 145
notched endurance limit, 75, 77, 79, 95, 116, 117, 119, 120, 125, 130, 132, 138, 139, 141, 145, 146, 152-154, 160, 162
notches, 30, 64, 77, 79, 96, 98, 125, 128, 130, 133, 142, 145, 154, 155, 161, 163, 168, 169

octrahedral shear stress compressive mean stress, 109, 144
Ono, 51, 55, 86, 98, 113-116, 162
opening mode, 39, 41, 42, 57, 72, 73
opening mode crack, 39, 41
opening mode propagation, 39
oxide layer, 167, 168

paired endurance limits, 95
parabolic stress wave traces, 65
photoelasticity, 65
plane-bending fatigue, 36, 93, 100
plane-strain failure, 64
plastic, 33-35, 38, 58, 62, 65, 73, 77, 101-103, 109, 134, 167, 168
plastic deformation, 33, 58, 77, 109, 134, 168
Poisson's ratio, 63
polished chamfer, 168
porosity, 63
pressure cone, 177-179
pressure cone angle, 177, 178
pre-tensioned, 171, 173, 174, 182
probability of fatigue, 5
profile keyway, 125
proof stress, 176-178
propagate, 3, 12, 13, 39, 41, 50, 51, 54, 55, 57, 60, 63, 65, 72, 99, 133, 159, 163, 169
pseudo factor of safety, 177
pure bending, 106
pure cosign curve, 85
pure torsion, 106

radial cracks, 55
random fatigue, 78
range of cyclic stress, 3
rear axle, 64
regression, 97, 98
relay can, 158, 159
reliability factor, 79
retardation, 96, 97

Index   **187**

rigid body analysis, 181
Rockwell C, 69, 161
rotating bending endurance limit, 34, 35, 134
rotating bending s_a-N curve, 33, 44, 45, 95, 119, 122, 123
rotating-bending fatigue, 1, 2, 33, 85, 114
Rotscher, 177
Runout, 6, 13, 32, 53, 56, 96, 97, 119

safe-life design, 82, 86
s_a-fnc experiment, 16
s_a-fnc test program, 16
s_a-N curve, 1, 2, 5, 33, 35, 43-45, 75, 81, 83, 91-93, 95-100, 104, 114, 119, 121-124, 166
scales, 65
scanning electron microscope, 41, 60, 61, 67
scatter of replicate s_a-N curves, 43
scoring, 166-168
scoring failures, 168
season cracking, 58
SEM micrographs, 61, 67, 69, 102
service fatigue failures, 47
service stress, 47
shank, 50, 171, 176
shape, 11, 13, 17, 68, 78, 93, 99, 104, 125, 154-156, 164, 171, 173
shape effect, 99
sharp corner edge, 99, 163
sharp notch, 103
shear failures, 70
shear lips, 50, 64, 110, 133
shear stress, 39, 85, 103, 106-110, 114-116, 122, 141, 143-148, 150, 171
shot-peening, 164, 167
sibling failures, 52
Sines, 109, 110
Sines' model, 109, 110
size effect, 75-77, 99
sizing, 37, 44, 139, 144, 148, 151
slant fracture crack, 53
sled-runner keyway, 125
sliding deformation, 57
sliding mode of failure, 68
slip, 29, 40, 83, 110

"Smith, J. O.", 127, 141
Snedecor, 103
Speed and Wave form effect, 78
speed effect, 78
splined shaft, 52
spotface, 168
spring rate, 157, 176-178, 180
square corners, 4
square fracture crack, 53
standard test specimen, 34, 75
stand-by state of stress (SSS), 144
static bending failure, 63
static equilibrium, 173
static strength, 76
static tensile yield, 26, 27, 35, 101, 106, 116, 128, 129
static tensile yield strength, 26, 27, 35, 101, 106, 116, 128, 129
static torsional overload failures, 59
static yield strength, 29, 33, 38, 76, 116, 127, 128
static yielding, 57
statistical distribution, 6, 43, 121
statically-equivalent bolt force, 6, 43, 121
steady torsional shear stress, 114-116
steady-state rotating bending, 113
step, 37, 88, 107
stop marks, 47-51, 55, 57
Straight-line segment, 6, 15-17, 124
strain-controlled bending, 105, 148
strain-controlled fatigue test, 26, 38, 82, 88, 97
strain-controlled test machine, 137
strain-gage extensometer, 101
strain-gage-based data, 29
stress relieving notches, 154, 155
strength, 6-8, 10, 11, 14-18, 20, 23-27, 29, 30, 33-35, 38, 76, 79, 83, 95, 96, 101, 102, 104, 106, 116, 120, 124, 127-130, 134, 135, 137, 141, 143, 145, 152, 153, 160-162, 166, 169, 176
strength distribution, 6, 16-18, 20, 161, 162
stress concentration, 3, 4, 12, 34, 48, 54, 77, 104, 116, 119-121, 130, 138, 151, 153-157, 162, 163, 165

stress concentration factor, 34, 77, 96, 104, 119-121, 130, 138, 151, 154-156
stress corrosion crack, 58
stress cycles, 1, 2, 4, 27, 33, 35, 42, 95, 122
stress redistribution, 29, 31, 35, 36, 131
stress relieving notches, 154, 155
stress reversal, 101
stress wave traces, 65
stress-strain hysteresis loops, 38, 101
striations, 41, 42, 73
Stulen, 83
surface finish, 17, 75, 77, 78, 98, 100, 163, 167, 168
surface finish effect, 77
surface grains, 29
surface hardening, 163, 166-168
surface metal grains, 29

tapered press-fit specimen, 3
tapered spindle, 51
temperature effect, 78
tensile failure, 71
tensile mean stress, 17, 110, 134, 159,
tensile strengths, 8, 77
tensile ultimate strength, 6, 17, 23-25, 34, 35, 76, 79, 95, 120, 122, 123, 132, 135, 145, 160, 162, 168
tension nut, 158, 180, 181
termination, 54
thread force distribution, 180, 181
threshold value, 6
torque-controlled, 141, 142
torsion, 1, 52-54, 58, 59, 71, 76, 83, 105-110, 113-117, 125, 127, 128, 141, 142

147, 148, 150, 160, 161, 171,
torsional fatigue failure, 52-54
torsional shear stress, 114-116
torsional stress, 109, 110, 115-117, 128, 141
transmission shaft, 53
transverse shear stress, 171
Type I censoring, 6

ugly s_a-N curves, 91, 121, 124
ultimate strength, 36, 76, 106, 145
unnotched endurance limit, 75, 77, 95, 119, 120, 130, 141, 153, 154, 160
unnotched specimens, 98, 101, 135, 137, 141
up-and-down test method, 16

v-slant failure, 64

Wallner lines, 63
wear failure, 47
Weibull life distribution, 74
"Wohler, August", 1
Wohler's test, 1, 4
Wood, 40, 56, 62
wood handle, 62
Woodruff keyway, 56, 125
wrought iron, 1-6, 17

X-ray analysis, 29

yielding fatigue, 92

zero velocity impact, 23

Author Biography

ROBERT EUGENE LITTLE (1933-2019) was a Professor of Mechanical Engineering at The University of Michigan-Dearborn for over 54 years. Professor Little received two B.S.E. degrees from The University of Michigan in 1959, an M.S.E. degree from The Ohio State University in 1960, and a Ph.D. degree from the University of Michigan in 1963. Professor Little published numerous books including, *Statistical Design of Fatigue Experiments, Manual on the Statistical Planning and Analysis of Fatigue Tests, Probability and Statistics for Engineers, Statistical Analysis of Fatigue Data, and Mechanical Reliability Improvement: Probability and Statistics for Experimental Testing,* with another book coming soon about statistical analysis of s-N data and Weibull distributions. Professor Little was a member of the American Society for Engineering Education (ASEE), the American Society for Testing and Materials (ASTM), the American Statistical Association (ASA), Pi Tau Sigma, Tau Beta Pi, Sigma Xi, and Phi Kappa Phi. His writings have been included in numerous publications including, ASTM Materials Research and Standards, ASTM STP, ASTM Journal of Materials, Journal of the American Statistical Association, ASTM Journal of Testing and Evaluation, The Aeronautical Quarterly, Technometrics, ASTM Annual Book of Standards, The Basic Fluid Power Research Journal, and Machine Design. Professor Little was an active consultant for machine design and product liability, dedicating significant effort to improving BB gun safety.